FUNDAMENTAL CONCEPTS IN

PHYSIOLOGY

AN ILLUSTRATIVE STUDY

FUNDAMENTAL CONCEPTS IN

PHYSIOLOGY

AN ILLUSTRATIVE STUDY

REVISED FIRST EDITION

BY CHI-MING HAI
BROWN UNIVERSITY

Bassim Hamadeh, CEO and Publisher
Carrie Montoya, Manager, Revisions and Author Care
Kaela Martin, Project Editor
Alia Bales, Production Editor
Emely Villavicencio, Senior Graphic Designer
Stephanie Kohl, Licensing Coordinator
Sean Adams and Allie Kiekhofer, Interior Designers
Natalie Piccotti, Director of Marketing
Kassie Graves, Vice President of Editorial
Jamie Giganti, Director of Academic Publishing

Cover image copyright © by Depositphotos / shmeljov, copyright © by Depositphotos / shmeljov.

Printed in the United States of America.

ISBN: 978-1-5165-2809-7 (pbk) / 978-1-5165-2810-3 (br)

cognella® | ACADEMIC PUBLISHING

BRIEF CONTENTS

DETAILED CONTENTS

INTRODUCTION

Human physiology is the study of how organ systems function interactively to support life. As a biological discipline, physiology is unique in having a focus on multiple layers of organization—organ systems, tissues, cells, and molecules. For example, cardiovascular physiology studies the function of the whole heart, cardiac muscle cells, ion channels, contractile proteins, neurotransmitters, and hormones. Physiology is the foundation of medicine. For example, cardiovascular physiology provides the foundation for understanding the causes of hypertension and heart failure. Respiratory physiology provides the foundation for understanding the causes of breathing difficulties in premature babies and asthmatic patients.

Physiology has been my intellectual home since my undergraduate years. I have three degrees in physiology, did my postdoctoral training in physiology, and have been teaching physiology at Brown University for more than twenty-six years. I did my undergraduate degree work in physiology at the University of Toronto, where I gained my first research experience in physiology as a summer research intern. I moved a bit north for work on my master's degree in physiology at the University of Ottawa, where I published my first paper in a physiology journal. I then turned south for my doctorate in the Department of Physiology at Johns Hopkins Medical School, where I began my long and continuing journey in smooth muscle research. After Hopkins, I did my postdoctoral research training in Dr. Richard Murphy's laboratory in the Department of Physiology at the University of Virginia. Dr. Murphy and I developed the four-state crossbridge model for explaining the regulation of smooth muscle contraction by myosin light chain phosphorylation. This model has had some influence in smooth muscle research. After Virginia, I joined the Department of Physiology and Biophysics at Brown University in 1988. After two promotions and twenty-six years, I am still conducting research on smooth muscle and teaching physiology at Brown. This book is a summary of the lectures I teach in a one-semester, intermediate-level undergraduate course in physiology, designed to prepare students for careers in a host of health-related disciplines, including medicine, physician assistant graduate programs, and biomedical engineering.

We live in a world of the Internet, where information is shared by people of the world. Many high-quality diagrams of physiology and medicine are now available online through both Wikimedia Commons and scientific journals. Authors of these diagrams are typically established scientists in their field of expertise. In the spirit of resource sharing, I have decided to use many diagrams from Wikimedia Commons and scientific journals, with permission from the publishers. And when not able to find the necessary diagrams on these resources, I created some diagrams myself.

I sincerely thank Dr. George Biro at the University of Toronto, Dr. Thomas Eddinger at Marquette University, and Dr. Hae Won Kim, and Mrs. Andrea Sobieraj at Brown University for reviewing some chapters of this book. Dr. Biro was a faculty member on my MS thesis committee at the University of Ottawa. Dr. Eddinger and I both did our postdoctoral training in Dr. Richard Murphy's lab at the University of Virginia. Dr. Kim is my colleague at Brown. Mrs. Sobieraj has been my senior teaching associate at Brown for longer than I can remember.

FUNDAMENTAL CONCEPTS IN PHYSIOLOGY

LEARNING OBJECTIVES

1. **Internal Environment.** Define and provide examples of internal environment.
2. **Homeostasis.** Define and provide examples of homeostasis.

WHAT IS HUMAN PHYSIOLOGY?

Physiology is about life—the function of living organisms. Human physiology is the study of how multiple organ systems enable a human being to live a healthy life. Diagnosis and treatment of diseases of our organ systems have been developed based on a clear understanding of physiology of organ systems at multiple layers of organization—molecules, cells, organ systems, and whole organism. This multilayer approach to physiology is emphasized in this book. Physiology is a quantitative biomedical science. Many major physiological concepts include quantitative analysis of vital variables, such as blood pressure, blood oxygen tension, blood glucose concentration, pH, etc. Multiple organ systems are often involved in the regulation of a single variable. Within limits, moderate dysfunction in one organ system can be compensated by adjustment in another organ system. For example, the lungs and kidneys together regulate plasma pH. In uncontrolled diabetes, the detrimental effect of excess acid production from dysfunctional fat metabolism on plasma pH can be minimized by increasing excretion of carbon dioxide by the lungs. An integrative understanding of the function of multiple organ systems is important for learning physiology.

FUNDAMENTAL CONCEPTS IN PHYSIOLOGY: INTERNAL ENVIRONMENT AND HOMEOSTASIS

Claude Bernard, a French physiologist, recognized the importance of a physiological "**internal environment**" for supporting the survival and function of all cells within the human body. Internal environment refers to the physical and chemical states of the extracellular fluid surrounding cells—temperature, oxygen tension, pH, etc. For example, the core temperature in a healthy individual is regulated at 37°C, with relatively small fluctuations during a day and menstrual cycle. Similarly, plasma pH in a healthy person is normally regulated at 7.4 by the coordinated action of the respiratory and kidney systems. Therefore, from a physiologist's point of view, the primary function of organ systems is to maintain a normal composition of the internal environment—oxygen, carbon dioxide, glucose, pH, etc.

Walter Cannon, an American physiologist, introduced the concept of **homeostasis** to describe the dynamic but stable steady states (blood pressure, oxygen, glucose, etc.) of a living human body that are maintained by a system of complex and coordinated physiological reactions. The term *homeostasis* is still being used to describe the normal steady states of physiological variables that are dynamically regulated by multiple and sometimes antagonistic regulatory mechanisms. For example, blood pressure homeostasis is regulated by the coordinated regulation of two branches of the autonomic nervous system—parasympathetic and sympathetic nervous systems. Similarly, plasma glucose homeostasis is regulated by insulin, glucagon, cortisol, and other hormones. Homeostasis is a useful concept for thinking about the healthy human body as integrated homeostasis of all physiological variables, when all physiological systems are operating optimally.

PHYSIOLOGICAL SYSTEMS

Prior to the creation of neuroscience as a scientific discipline, physiology covered all organs of the human body, including the nervous system as neurophysiology. Neuroscience is now typically taught as an independent course in most colleges and universities. To avoid duplication, I cover mostly non-neural organ systems in this book. Exceptions include neural control of muscle contraction and the autonomic nervous system. The autonomic nervous system regulates almost all organ systems. The following summarizes the topics covered by the chapters in this book.

Cell Membrane Transport and Signaling (Chapter 2). Cell membrane is the interface between the inside and outside of a cell. The cell membrane regulates the transport between intracellular and extracellular fluids, a fundamental process for the function of all organ systems—for example, transport of oxygen and carbon dioxide in the respiratory system and transport of ions and water in the renal system. In addition, specialized protein molecules (receptors) located in the cell membrane can respond to specific extracellular signal molecules (hormones and neurotransmitters) by activating intracellular processes, leading to specific cell functions, such as secretion and contraction.

Cell Membrane Potentials, Synapses, and Autonomic Nervous System (Chapter 3). Selective transport of electrically charged ions across the cell membrane results in an electrical potential across the cell membrane, which is the fundamental mechanism underlying the excitability of nerve and muscle cells. The coupling between receptor signaling and ion transport is the fundamental mechanism underlying synapses at neuron-neuron and neuromuscular junctions. The autonomic nervous system, consisting of parasympathetic and sympathetic branches, regulates the activity of almost all organ systems in response to whether a person is in the resting or fight-or-flight state.

Muscle Physiology (Chapter 4). Muscle contraction is perhaps the most explicit sign of life. The musculoskeletal system under the control of the somatic nervous system supports the posture and movement of the human body. Motor neurons from the somatic nervous system form junctions with skeletal muscle cells, where a neurotransmitter released by motor neurons activates receptors on skeletal muscle cells to initiate a cascade of intracellular signaling processes that lead to muscle contraction. Finally, biochemical and biophysical interactions between two proteins inside a muscle cell, actin and myosin, enable the conversion of biochemical energy from breaking a chemical bond to the production of mechanical energy in the form of molecular motion.

Endocrine Physiology (Chapter 5). Hormones are powerful molecules. At extremely low concentrations in the plasma, hormones regulate the function of multiple organ systems, such as the basal metabolic rate, plasma glucose metabolism, cardiac output, and menstrual cycle. Hormones have unique chemical structures that enable them to activate only specific receptors in specific target organs. The hypothalamus and pituitary gland in the brain contain seven major hormonal systems that regulate peripheral endocrine glands and other organ systems, including the thyroid gland, liver, kidney, and uterus. Insufficient or excessive secretion of a hormone can lead to diseases. For example, insufficient secretion of thyroid hormone in newborns can lead to irreversible retardation in brain development; therefore, newborns are routinely screened for thyroid insufficiency and, if necessary, treated with a thyroid hormone. In contrast, an excessive thyroid hormone secretion can lead to uncontrolled increases in heart rate and body metabolism. Some hormones—e.g., insulin and cortisol—are absolutely essential for survival. The availability of insulin now allows a person with type 1 diabetes mellitus to live a reasonably normal life. Type 2 diabetes mellitus, however, is increasingly becoming a major public health problem in affluent societies.

Cardiovascular Physiology (Chapter 6). Blood oxygen transport by the cardiovascular system is a major determinant of the level of physical activity that a person can achieve. Oxygen transport is a product of cardiac output (the rate of blood flow pumped by the heart through the circulatory system) multiplied by the blood oxygen content, which is determined by hemoglobin concentration and its oxygenation. Therefore, dysfunction of the cardiovascular system such as cardiac failure can lead to abrupt termination of physical activity, and even death. Recognizing the critical role of the cardiovascular system in determining performance in endurance exercise, some endurance athletes have illegally boosted blood oxygen transport by blood transfusion (blood doping) prior to a competition.

Respiratory Physiology (Chapter 7). Breathing is a telling sign of life. The American Lung Association used the following slogan some years ago: "When you can't breathe, nothing else matters." Breathing is a major challenge to asthmatic patients because their hyperactive airways can easily close. A limited list of medicine is available to dilate airways in asthmatic patients. Premature babies have breathing difficulties because immature lungs are stiff due to insufficient production of a biological molecule, surfactant, in the lungs. Synthetic surfactant is now available for improving the lung function of premature babies. High altitude is a challenging environment for the human body because the low atmospheric pressure results in a low partial pressure of oxygen for blood oxygenation. Adaptation to high altitude illustrates the important role of the respiratory system in regulating both blood oxygen content and pH.

Renal Physiology (Chapter 8). Kidneys are essential for removal of waste products from the body, activation of vitamin D, and production of erythropoietin. Patients with kidney failure need dialysis regularly to remove waste products from the body, and often erythropoietin and activated vitamin D supplementation for survival. The kidneys—by controlling the output from the body—and the gastrointestinal system—by controlling the input to the body—together regulate the chemical composition of the extracellular fluid. The kidneys consist of blood vessels and urinary tubules, where waste products are transported from the blood to the urine. Therefore, renal physiology is concerned with understanding the basic mechanisms for water and solute transport, as well as their regulation by several hormonal systems. Discovery of the transport mechanisms in the kidneys has led to the development of drugs for treating hypertension by controlling extracellular fluid volume.

Gastrointestinal Physiology and Regulation of Substrate Metabolism (Chapter 9). The ingestion, digestion, and absorption of food by the gastrointestinal system are essential for survival of the human body. Excessive intake of food, however, can result in obesity and associated diseases such as diabetes, atherosclerosis, and cancer. The gastrointestinal system is essentially a tube that begins in the mouth and ends in the anus. Movement of food along the gastrointestinal tract is powered mostly by smooth muscle cells and regulated by multiple sphincters. To extract nutrients from food, secretory cells lining the gastrointestinal system produce digestive enzymes and other factors for breaking down food into small molecules, which are then transported by absorptive cells to the blood and lymphatic circulation. Multiple hormones regulate the function of the gastrointestinal tract. Hormones and the autonomic nervous system regulate gastrointestinal motility and secretion. In addition, multiple hormones are involved in the control of appetite.

Reproductive Physiology (Chapter 10). Thanks to reproductive physiology, we have all arrived safely in this world. I find reproductive physiology a particularly fascinating subject because it not only regulates the health of the individual, but also enables the production of new human beings. Both male and female reproductive systems consist of reproductive organs for producing sperms or eggs, and endocrine organs for regulating the function of reproductive organs. There are major similarities and differences between the male and female reproductive systems. For example, testosterone is the major male reproductive hormone produced by the testis, whereas estrogen and progesterone are the major female reproductive hormones produced by the ovaries; however, both male and female endocrine systems utilize the same hypothalamic and pituitary hormones to regulate reproductive function. Furthermore, the female reproductive hormones oscillate during menstrual cycles, whereas the male reproductive hormones are relatively stable. Meiosis is the basic process for production of both sperms and oocytes; however, the progression of first and second meiosis in the male reproductive system is continuous, whereas the progression of second meiosis in the female reproductive system requires fertilization of the oocyte. Understanding reproductive physiology has led to the development of contraceptives and in vitro fertilization, which have had a dramatic effect on human population.

KEY TERMS

- homeostasis
- internal environment
- physiological systems

2

CELL MEMBRANE TRANSPORT AND SIGNALING

The cell membrane is a selective and regulated barrier between intracellular and extracelalular fluids. As shown in Fig. 2.1, a phospholipid bilayer forms the basic backbone of the cell membrane. Small and lipid-soluble molecules (e.g., oxygen and carbon dioxide) can cross the phospholipid bilayer by simple diffusion, whereas large and charged molecules require transport proteins in the cell membrane for crossing the cell membrane.

DIFFUSION

Diffusion is a process by which molecules move from a region of higher concentration to a region of lower concentration, as shown in Fig. 2.2. Random motion of molecules is the fundamental mechanism of diffusion. Molecules in the region of higher concentration have a higher probability of molecular collisions, which can result in movement to the region of lower concentration. In comparison, molecules in the region of lower concentration have a lower probability of molecular collisions. Concentration difference is the fundamental driving force for diffusional transport. For diffusion of a molecule across the cell membrane, the permeability of the cell membrane for the molecule and the total surface

LEARNING OBJECTIVES

1. **Diffusion.** Discuss the basic mechanism of diffusion; illustrate the calculation of net flux of diffusion from concentration gradient, permeability coefficient and surface area; compare and contrast simple diffusion and facilitated diffusion in terms of mechanisms and function; provide examples of simple and facilitated diffusion.

2. **Osmosis.** Define osmosis, and explain how osmotic pressure and hydrostatic pressure together determine net movement of water across a semi-permeable membrane.

3. **Osmolarity.** Explain the calculation of osmolarity from molarity of dissociable and/or non-dissociable molecules.

4. **Ion Channels.** Compare and contrast the three classes of ion channels in terms of gating characteristics and physiological function.

5. **Carrier-Mediated Transport.** Compare and contrast the three classes of carrier-mediated transport in terms of defining characteristics and physiological function, and provide examples of each class of carrier-mediated transport.

6. **Epithelial Transport.** Explain how transporters at the basolateral side and luminal side of intestinal epithelial cells mediate intestinal absorption of glucose.

7. **G-Protein-Coupled Receptors (GPCRs).** Discuss the signaling pathways of G-protein-coupled receptors, and provide an example of GPCR.

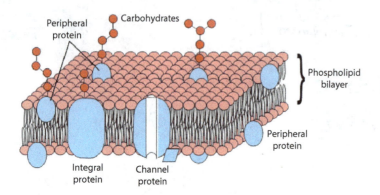

Fig. 2.1—Basic Structure of Cell Membrane. The cell membrane is a regulated barrier between the inside and outside of a cell. The phospholipid bilayer, major component of the cell membrane, is permeable to small and lipid-soluble molecules such as oxygen, carbon dioxide, and water. Membrane transport proteins—ion channels and carriers—are necessary for the transport of charged ions and large molecules across the cell membrane.

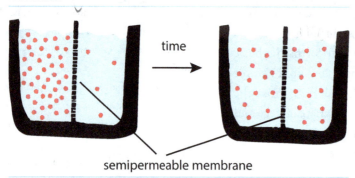

Fig. 2.2—Diffusion Across a Membrane. In this example, the semi-permeable membrane is permeable to the red molecules, allowing the net movement of the molecules from the compartment of higher concentration to the lower concentration. After sufficient time for diffusion, concentrations of the red molecule in the two compartments become equal.

area available for diffusion are additional determinants of diffusional transport. The following equation summarizes the contributions of these variables to the net flux of diffusion in moles/time:

$$\text{Net Flux} = \text{Permeability} \times \text{Surface Area} \times (\text{Concentration}_1 - \text{Concentration}_2)$$

Diffusional transport of O_2 and CO_2 between air and blood in the lungs is highly efficient, because permeability of the cell membrane to O_2 and CO_2 is high, total surface area for diffusion is enormous, and concentration difference between air and blood is large.

OSMOSIS—DIFFUSION OF WATER

Osmosis describes the diffusion of water from a region of higher water concentration to a region of lower water concentration. Diffusion of water is important in physiology because water is the major constituent of cells. Abnormal cellular gain of water leads to cell swelling, and abnormal cellular loss of water leads to cell shrinking.

OSMOLARITY—INVERSE MEASURE OF WATER CONCENTRATION

Osmolarity is the concentration of solute particles in a solution. Solute particles include all molecules and ions that can compete with water molecules for diffusion from one region to another. For example, Na^+, K^+, Cl^-, and glucose are solute particles in the extracellular fluid.

Osmolarity is an inverse measure of water concentration in the sense that high osmolarity implies low water concentration and low osmolarity implies high water concentration. Osmolarity, instead of water concentration, is used to predict the diffusion of water—and for two major reasons. First, calculating solute concentration instead of water concentration is a more sensitive method for predicting the direction of water movement, because the concentration of solute is much lower than the concentration of water in body fluids. In the extracellular fluid, for instance, concentration of the major cation Na^+ is only 0.14 M, whereas concentration of water is more than 55 M. Cell volume is highly sensitive to change in solute concentration. For example, a change in extracellular $[Na^+]$ from 0.14 M to 0.10 M, associated with a relatively small change in water concentration, is sufficient to cause significant cell swelling. Second, normal osmolarity of extracellular fluid serves as a reference value for preparing physiological solutions for intravenous infusion. For example, 0.9% NaCl and 5% glucose solutions are similar in osmolarity and commonly used for intravenous infusions.

Osmolarity of a solution can be calculated using the following equation:

$$\text{Osmolarity} = \text{Molarity} \times \text{Number of Solute Particles/Molecule}$$

For non-dissociable molecules, the number of solute particles/molecule equals 1, and osmolarity equals molarity, as shown in the following example:

$$\text{Osmolarity of 1 mM Glucose} = 1 \text{ mOsM (1 millimoles of solute particles/liter)}$$

For dissociable molecules, the number of solute particles/molecules equals the number of dissociated ions/molecule, as shown in the following two examples:

NaCl dissociates into $Na^+ + Cl^-$ in solution—that is, two solute particles/mole NaCl.

$$\text{Osmolarity of 1 mM NaCl} = 2 \text{ mOsM (2 millimoles of solute particles/liter)}$$

$MgCl_2$ dissociates into $Mg + 2Cl^-$ in solution—that is, three solute particles/mole $MgCl_2$.

$$\text{Osmolarity of 1 mM } MgCl_2 = 3 \text{ OsM (3 millimoles of particles/liter)}$$

As cited previously, osmolarity is an inverse measure of water concentration—that is, high osmolarity implies low water concentration and low osmolarity implies high water concentration. Accordingly, when two solutions of different osmolarity are separated by a semi-permeable membrane that is exclusively permeable to water, the direction of net movement water is expected to be from a solution having lower osmolarity to a solution having higher osmolarity. For example, when a solution having 1 mM NaCl is separated from a solution having 1 mM $MgCl_2$, the direction of water movement is expected to be from the NaCl solution to $MgCl_2$ solution.

OSMOTIC PRESSURE

Osmotic Pressure of a solution having nonzero osmolarity is the potential hydrostatic pressure that can be established by the diffusion of water into the solution. As illustrated in Fig. 2.3 (**left panel**), a solution having nonzero osmolarity held inside a glass column is separated from pure water outside the glass column by a semi-permeable membrane that is exclusively permeable to water. Initially, as shown in Fig. 2.3 (**left panel**), the levels of solution inside and outside the glass column were identical. At equilibrium, as shown

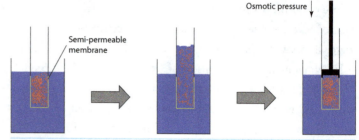

Fig. 2.3—Osmotic Pressure. Osmotic pressure of a solution containing an impermeable solute is the pressure that can be exerted by the diffusion of water into the solution. In the left panel, a tube containing an impermeable solute is separated from the surrounding water by a semi-permeable membrane that is exclusively permeable to water. In the middle panel, the diffusion of water into the solute-containing solution in the tube causes the solution in the tube to rise above the water level outside the tube. In the right panel, the application of hydrostatic pressure at exactly the osmotic pressure opposes the diffusion of water into the tube, thereby holding the water level inside the tube to be the same as the water level outside the tube.

in Fig. 2.3 (**middle panel**), the diffusion of water into the solution causes the level of solution inside the glass column to rise above the level of solution outside the glass column. An equilibrium is established when the outward movement of water driven by hydrostatic pressure equals the inward movement of water driven by osmotic pressure. The hydrostatic pressure established at equilibrium is named osmotic pressure, which is proportional to the osmolarity of impermeable solutes in the solution. As shown in Fig. 2.3 (**right panel**), the diffusion of water driven by osmotic pressure can be prevented by exerting hydrostatic pressure onto the solution inside the glass column at exactly the same level as osmotic pressure. In reverse osmosis for water purification, it is possible to extract pure water from a solution containing nonzero osmolarity through a semi-permeable membrane by exerting hydrostatic pressure that is higher than the osmotic pressure. As illustrated in the above example, hydrostatic and osmotic pressures must be considered together for predicting the direction of water movement.

It is important to note that osmotic pressure is calculated based on impermeable solute only, because osmotic pressure can be established only if the solute cannot permeate the membrane. Permeable solutes do not contribute to osmotic pressure, because the diffusion of permeable solute across the membrane will equalize and eliminate the osmotic gradient for net water movement.

DIFFUSION OF IONS THROUGH PROTEIN PORES (ION CHANNELS)

The phospholipid bilayer of the cell membrane is impermeable to charged molecules—e.g., ions such as Ca^{2+}, Na^+, and Cl^-. The diffusion of ions through ion channels is the basic mechanism of electrical excitability in electrically excitable cells—neurons and muscle cells. Ion channels are macromolecular protein complexes that form ion-selective pores in the cell membrane for the diffusion of specific ions. Each ion channel is made up of hydrophobic domains that anchor the channel in the phospholipid bilayer of the cell membrane and other domains that form the channel pore. Ion channels are typically selectively permeable to an individual or class of ion(s). For example, a Na^+ channel is selectively permeable to Na^+, and a K^+ channel is selectively permeable to K^+. A nonspecific cation channel is permeable to both Na^+ and K^+, but not Cl^-. Charge groups and pore size of ion channels are important determinants of channel selectivity.

Gating of Ion Channels. Ion channels can exist in an open or a closed state. The open state of an ion channel is permeable to ions, whereas the

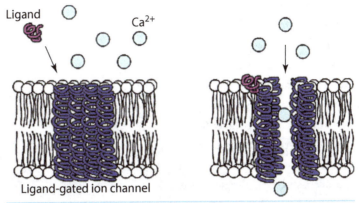

Fig. 2.4—Diffusion of Ions Through a Ligand-gated Ion Channel. The phospholipid bilayer of the cell membrane is impermeable to charged molecules. Ion channels are membrane proteins having a pore that allows the diffusion of specific ions between extracellular and intracellular fluids. The opening and closing of the pore of ion channels are regulated by various gating mechanisms. This example shows the regulation of a ligand-gated Ca^{2+} channel. The binding of a ligand to the receptor site of this ion channel induces the opening of the channel for the diffusion of Ca^{2+} down its concentration gradient.

closed state is impermeable to ions. The transition between open and closed states of an ion channel is known as gating. Ion channels can be classified based on gating mechanisms. As shown in Fig. 2.4, a ligand-gated ion channel will increase its open probability in response to binding of a ligand to the receptor domain of the channel. Ligand-gated channels are also known as receptor ion channels. For example,

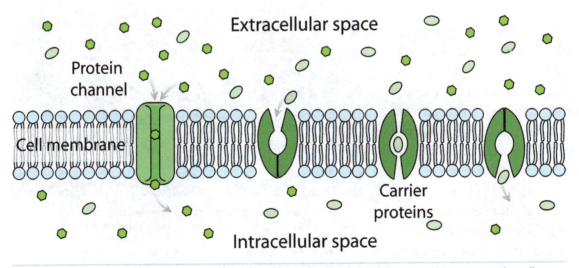

Fig. 2.5—Carrier-mediated Facilitated Diffusion. The diffusion of large molecules such as glucose across the cell membrane is mediated by a carrier protein. Unlike ion channels, which contain a pore for ionic diffusion, a carrier protein binds to a substrate on one side of the cell membrane and the substrate-carrier complex then undergoes conformational change to transport the substrate down its concentration gradient to the other side of the cell membrane.

at the neuromuscular junction, the neurotransmitter acetylcholine (a ligand) binds to acetylcholine receptor ion channels on the skeletal muscle cell membrane to cause an increase in the open probability of the channel. A voltage-gated ion channel will increase its open probability in response to a change in voltage across the cell membrane. Opening and closing of voltage-gated Na^+ and K^+ channels are the fundamental mechanisms of impulse generation in neurons. A stretch-activated ion channel increases its open probability in response to mechanical stretch on the channel. Stretch-activated ion channels are the fundamental mechanisms of sound detection by hair cells in the inner ear.

CARRIER-MEDIATED TRANSPORT

The phospholipid bilayer of the cell membrane is impermeable to large molecules—for example, glucose. Facilitated diffusion is a mechanism in which a carrier protein in the cell membrane transports a molecule down its concentration gradient. A carrier protein for facilitated diffusion is a membrane protein that has a binding site for a substrate and is capable of undergoing conformational change for transporting the bound substrate from one side of the cell membrane to the other side. As shown in Fig. 2.5, a substrate molecule is bound to the carrier

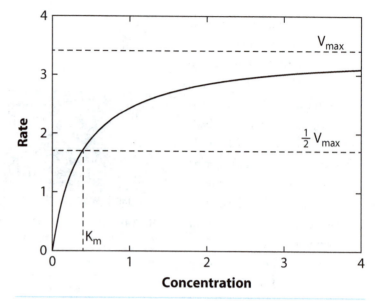

Fig. 2.6—Saturation Kinetics of Carrier-mediated Transport. Binding of a substrate to a carrier is the rate-limiting step in carrier-mediated transport. Increasing substrate concentration will increase the transport rate up to the maximum rate (V_{max}), when all carriers are bound (saturated) by the substrate. K_m is the substrate concentration at 50% V_{max}, when 50% of carriers is bound by the substrate.

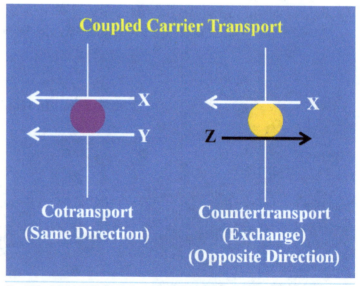

Fig. 2.7—Coupled Carrier Transport. Carrier-mediated transporters are capable of transporting two substrates together either in the same direction (cotransport) or in the opposite direction (countertransport or exchange). Coupled carrier transport is often considered secondary active transport, because the concentration gradient of one substrate can be used to drive the transport of another substrate against its concentration gradient. For example, Na^+-glucose cotransporters on intestinal epithelial cells utilize the Na^+ gradient to drive the transport of glucose from a lower [glucose] in the intestinal lumen to the higher [glucose] inside epithelial cells.

protein facing the extracellular space. The carrier then undergoes a conformational change that causes the release of the bound substrate to the intracellular space. The direction of net movement of substrate via carrier-mediated facilitated diffusion will be from higher concentration to lower concentration, because the probability of carrier-substrate formation is proportional to substrate concentration.

As shown in Fig. 2.6, carrier-mediated facilitated diffusion exhibits **saturation kinetics**. The rate of facilitated diffusion approaches a maximum (V_{max}) with an increase in substrate concentration. V_{max} reflects the total transport capacity of carrier proteins when all binding sites of carrier proteins are occupied by the substrate. K_m—the substrate concentration at which transport rate is 50% V_{max}—reflects the binding affinity of the carrier for substrate. A high K_m indicates low binding affinity, whereas a low K_m indicates high binding affinity.

Coupled Carrier Transport. Coupled carrier transport mediates the transport of two substrate molecules simultaneously. As shown in Fig. 2.7 (**left panel**), a coupled carrier transport is named cotransport, when the carrier transports two molecules in the same direction. For example, Na^+-glucose cotransport mediates the simultaneous transport of Na^+ and glucose by intestinal epithelial cells. As shown in Fig. 2.7 (**right panel**), a coupled carrier is named countertransport or exchange, when the carrier transports two molecules in the opposite direction. For example, Na^+/Ca^{2+} exchange mediates the transport of Na^+ into and Ca^{2+} out of cardiac muscle cells.

Coupled carrier transport is often considered secondary active transport, because the concentration gradient of one substrate is often used to drive the transport of the other substrate against its concentration gradient. For example, during intestinal absorption of glucose, Na^+-glucose cotransport utilizes Na^+ gradient to drive the uptake of glucose by intestinal epithelial cells, which results in near zero [glucose] in the intestinal lumen and high [glucose] inside intestinal epithelial cells.

Primary Active Transport. As shown in Fig. 2.8, primary active transport proteins are enzyme transporters capable of hydrolyzing ATP to produce energy for the transport of a molecule against its concentration gradient. A primary active transporter is named based on the transporter molecule (X) as X-ATPase. For example, Ca^{2+}-ATPase hydrolyzes ATP to produce energy for Ca^{2+} extrusion

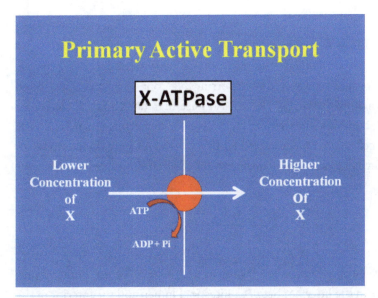

Fig. 2.8—Primary Active Transport. Primary active transport is named after the substrate (X) it transports as X-ATPase. Primary active transport directly utilizes energy from ATP hydrolysis to transport substrate X against its concentration gradient. For example, Na^+-K^+-ATPase utilizes energy from ATP hydrolysis to transport of Na^+ out of a cell and K^+ into a cell against their concentration gradients, thereby maintaining a low $[Na^+]$ and high $[K^+]$ inside a cell.

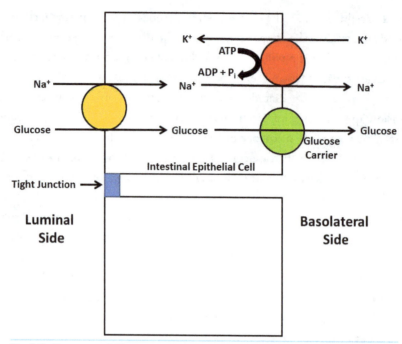

Fig. 2.9—Transport of Glucose and Na⁺ by Intestinal Epithelial Cells. On the luminal membrane of intestinal epithelial cells, glucose is transported from lumen into intestinal epithelial cells by Na⁺-glucose cotransport. At the basolateral membrane of intestinal epithelial cells, glucose is transported out of the cell to capillary blood by facilitated diffusion, and Na⁺ is transported out of the cell to capillary blood by Na⁺-K⁺-ATPase.

by cardiac muscle cells. Similarly, H⁺-ATPase hydrolyzes ATP to produce energy for acid excretion by renal tubular cells. Some primary transporters mediate the transport of two molecules. For example, Na⁺-K⁺-ATPase hydrolyzes ATP to produce energy for transporting Na⁺ out of and K⁺ into a cell against their concentration gradients. In almost all cells, Na⁺-K⁺-ATPase maintains a low intracellular [Na⁺] against a high extracellular [Na⁺] and high intracellular [K⁺] against a low extracellular [K⁺]. The Na⁺ concentration gradient across the cell membrane, established by Na⁺-K⁺-ATPase, functions as an energy source for cotransporters and exchangers to drive the transport of other molecules against their concentration gradients.

EPITHELIAL TRANSPORT

Epithelial cells form the interface between the external environment and interior of the human body. For example, intestinal epithelial cells form the interface between intestinal lumen and the interior of the human body. Airway epithelial cells form the interface between lung alveoli (air sacs) and the interior of the human body. All epithelial cells have a luminal surface facing the external environment and a basolateral side facing the internal environment. As shown in Fig. 2.9, different transporters located at the luminal and basolateral sides of intestinal epithelial cells function together to mediate intestinal

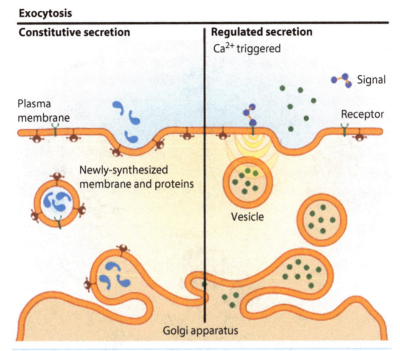

Fig. 2.10—Exocytosis. Exocytosis serves the function of: a) releasing contents of membrane vesicles to the extracellular space—e.g. release of neurotransmitter by nerve terminals, and b) insertion of membrane-bound proteins onto the cell membrane—e.g. insertion of glucose transporters onto the cell membrane in response to insulin. The process of exocytosis may be constitutively active or regulated by extracellular signals.

absorption of glucose. The Na⁺-glucose cotransport, on the luminal side of intestinal cells, utilizes Na⁺ gradient to transport glucose against its concentration gradient into intestinal epithelial cells, resulting in low luminal [glucose] and high intracellular [glucose] in intestinal epithelial cells. The high [glucose] inside intestinal epithelial cells then drives the movement of glucose into blood by facilitated diffusion via the glucose carrier on the basolateral side.

EXOCYTOSIS OF MEMBRANE VESICLES

Exocytosis is a cellular export mechanism by which a cell packages molecules within the content or on the membrane of intracellular membrane vesicles for fusion with the cell membrane. As shown in Fig. 2.10, molecules within the content of membrane vesicles are exported to the extracellular fluid; for example, neurons package neurotransmitters inside synaptic vesicles for exocytosis into the synaptic cleft. In comparison, molecules packaged on the membrane of membrane vesicles are inserted into the cell membrane. For example, insulin stimulates glucose uptake by muscle cells by regulating the fusion of intracellular membrane vesicles containing the glucose carrier GLUT-4 on the vesicular membrane with the cell membrane, resulting in the insertion of GLUT-4 onto the muscle cell membrane.

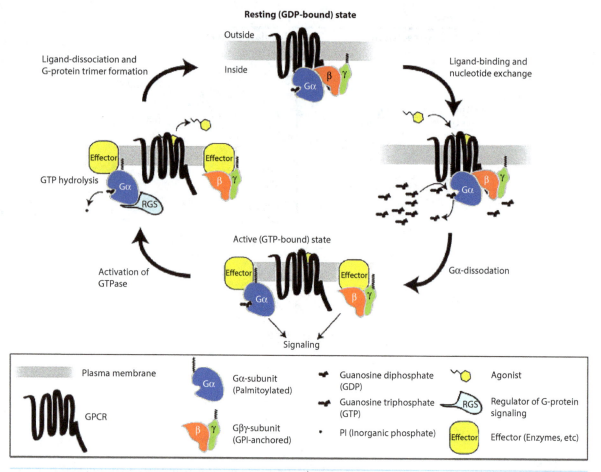

Fig. 2.11—G-Protein-coupled Receptor (GPCR) Signaling Mechanism. GPCR mechanism is commonly used by many cell types to transduce an external signal (ligand) into an intracellular response. As shown at the top of this figure, the ligand-free resting state of the GPCR complex consists of a membrane receptor and three tightly associated G-protein subunits (α, β, γ) on the inside of a cell membrane. In this resting state, the α subunit is bound by GDP. Binding of a ligand to the receptor induces conformational change of the GPCR complex, resulting in the exchange of GTP with GDP on the α subunit. The active GTP-bound α subunit then dissociates from the GPCR complex to activate an effector enzyme—e.g. adenylate cyclase, phospholipase C, etc.. The activated enzyme typically catalyzes the production of one or more intracellular messengers—e.g. cyclic AMP, inositol trisphosphate, diacylglycerol, etc. The α subunit is a GTPase, capable of hydrolyzing the bound GTP to GDP, thereby returning the α subunit to the resting GDP-bound state. Dissociation of the ligand from the receptor induces the re-association of the GDP-bound α subunit with the GPCR complex.

G-PROTEIN-COUPLED RECEPTOR SIGNALING MECHANISM

G-protein-coupled receptors are membrane receptors that utilize GTP-binding proteins for transducing the signal of receptor binding by a ligand to the activation of effector proteins; a G protein-coupled receptor is made up of a receptor and a dissociable GTP-binding protein complex consisting of α, β, γ subunits. The α-subunit—the GTP-binding subunit of the G protein complex—has a binding site for GDP or GTP. A GDP-bound α-subunit is inactive, whereas a GTP-bound α-subunit is active and capable of activating effector proteins. As shown in Fig. 2.11, a resting receptor is associated with a G protein complex having GDP bound to the α-subunit. Binding of a ligand to the receptor causes an exchange of GTP for GDP on the α-subunit and dissociation of the G protein complex from the receptor into the GTP-bound α-subunit and the $\beta\gamma$ subunit complex. The GTP-bound active α-subunit is capable of activating an effector enzyme for catalyzing the production of intracellular messengers for cell activation. Sometimes the $\beta\gamma$ subunit may also activate another effector enzyme for cell activation. The α-subunit is a GTPase capable of self-inactivation by hydrolyzing the bound GTP into GDP. The cycle of G protein activation and inactivation continues as along as a ligand is bound to the receptor. Removal of the ligand from the receptor would drive the cycle to G protein inactivation and reassociation of the G protein complex with the receptor.

In cardiac muscle cells, β-adrenergic receptors are G protein-coupled receptors that transduce the signal of receptor binding by the neurotransmitter norepinephrine to the strengthening of cardiac muscle contraction. Binding of norepinephrine to β-adrenergic receptors causes activation of the G protein G_s. The GTP-bound activated Gs then dissociate from the receptor to activate the effector enzyme, adenylate cyclase, which catalyzes the production of cyclic AMP for strengthening the contraction of cardiac muscle cells.

KEY TERMS

- carrier-mediated transport
- coupled carrier transport
- diffusion
- epithelial transport
- exocytosis

- facilitated diffusion
- G-protein-coupled receptor signaling
- gating of ion channels
- ion channels

- osmolarity
- osmosis
- osmotic pressure
- primary active transport
- saturation kinetics

IMAGE CREDITS

3

CELL MEMBRANE POTENTIALS, SYNAPSES, AND AUTONOMIC NERVOUS SYSTEM

LEARNING OBJECTIVES

1. **Electrochemical Equilibrium.** Define electrochemical equilibrium, and illustrate the calculation of Nernst Potential for monovalent and divalent cations and anions at a given concentration gradient.

2. **Resting Membrane Potential.** Explain the dependence of resting membrane potential on the concentration gradient and membrane permeability of Na^+ and K^+ conceptually and mathematically using the Goldman–Hodgkin–Katz equation.

3. **Graded Potential.** Explain the generation of graded potential by receptor ion channels, and discuss the characteristics of graded potentials.

4. **Action Potential.** Discuss the ionic basis of action potential in a typical neuron, and explain the ionic mechanisms underlying absolute and refractory periods.

5. **Chemical Synapse.** Describe the structure and function of a typical chemical synapse; define and provide examples of subthreshold and suprathreshold synapses; explain temporal and spatial summation of postsynaptic potentials at subthreshold synapses.

6. **Autonomic Nervous System.** Compare and contrast the two branches of autonomic nervous system in terms of anatomy, neurotransmitters, and function.

Most mammalian cells exhibit an electrical potential across the cell membrane, with the inside of the cell negative relative to the outside, as shown in Fig. 3.1. Cell membrane potentials are expressed as inside relative to the outside of a cell—for example, a typical membrane potential in quiescent neurons that are not actively releasing neurotransmitters is −70 mV. Cell membrane potential is an important determinant of the activity of excitable cells—neurons and muscle cells. A change in membrane potential toward a more positive value, for example, from −70 mV to +20 mV, is sufficient to stimulate neurotransmitter release by neurons and contraction of skeletal muscle cells. The presence of a cell membrane potential indicates a difference in electric charge across the cell membrane. Ions are carriers of electric charge in solution. Accordingly, cell membrane potential is established by the diffusion of ions; for example, Na^+, K^+, and Cl^-, through ion channels, as driven by their concentration gradients.

17

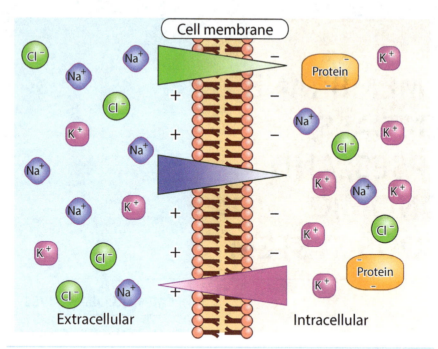

Fig. 3.1—Membrane Potential. Most mammalian cells exhibit a membrane potential that is negative inside relative to outside as a result of differential ionic gradients of Na$^+$, K$^+$, and Cl$^-$ across the cell membrane and the differential permeability of the cell membrane to these ions.

In the following sections, we first address the Nernst potential, an equilibrium potential established by electrochemical equilibrium for a single ion. We then look at the resting membrane potential, a steady-state potential established by the balance of multiple ionic fluxes across the cell membrane.

ELECTROCHEMICAL EQUILIBRIUM AND NERNST EQUATION

Nernst potential is an electrical potential established by electrochemical equilibrium for a single ion. Fig. 3.2 shows a hypothetical cell having a relatively higher concentration of a cation X$^+$ in the intracellular fluid than extracellular fluid. A conjugate anion is present at the same concentration as X$^+$ in both intracellular and extracellular fluids, but is not considered here, because the cell membrane is exclusively permeable to X$^+$. The concentration gradient of X$^+$ across the cell membrane is expected to cause the diffusion of X$^+$ out of the cell, causing the accumulation of negative charge inside the cell and positive charge outside the cell. The resulting electrical potential across the cell membrane will drive the diffusion of X$^+$ into the cell, thereby decreasing the net efflux of X$^+$. At electrochemical equilibrium, the efflux of X$^+$ driven by the concentration gradient is exactly balanced by the influx of X$^+$ driven by the electrical gradient. The membrane potential at electrochemical equilibrium is known as Nernst potential, which can be predicted by the following Nernst equation:

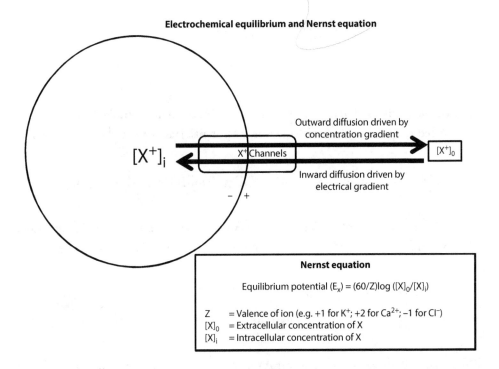

Fig. 3.2—Electrochemical Equilibrium and Nernst Potential. When there is an ionic concentration gradient across a membrane that is permeable to the ion, diffusion of ions from higher concentration to a lower concentration causes the development of a membrane potential (Nernst Potential). In this example, the concentration of a cation X^+ is higher inside a cell having a cell membrane that is exclusively permeable to X^+. Diffusion of X^+ down its concentration gradient through X^+ channels—from inside to the outside of the cell—causes the separation of charge and development of a membrane potential. At electrochemical equilibrium, the outward flux of X^+ driven by the concentration gradient equals the inward flux of X^+ driven by the electrical potential (Nernst potential).

$$\text{Nernst Potential (in mV)} = (60/Z)\log([X]_o/[X]_i)$$

$[X]_o$ and $[X]_i$ are the extracellular and intracellular concentrations of X, respectively. Z is the valence of the ion X—for example, the valence for Na^+ is +1; the valence for Cl^- is −1; and the valence for Ca^{2+} is +2.

The following examples illustrate the calculation of Nernst potentials for monovalent cation, monovalent anion, and divalent cation.

Example 1: Nernst Potential for Monovalent Cation. Consider the following extracellular and intracellular concentrations of KCl for a cell, whose cell membrane is exclusively permeable to K^+:

Extracellular fluid: 10 mM KCl

Intracellular fluid: 100 mM KCl

The diffusion of K^+, driven by its concentration gradient from intracellular fluid to the extracellular fluid, is expected to result in a negative membrane potential, as predicted by the Nernst equation.

$$\text{Nernst Potential for } K^+ = 60/(+1) \log(10\,\text{mM}/100\,\text{mM}) = -60\,\text{mV}$$

Example 2: Nernst Potential for Monovalent Anion. Consider the following intracellular and extracellular concentrations of NaCl for a cell, whose cell membrane is exclusively permeable to Cl^-:

Extracellular fluid: 100 mM NaCl

Intracellular fluid: 10 mM NaCl

The diffusion of Cl^-, driven by its concentration gradient from extracellular fluid to the intracellular fluid, is expected to result in a negative membrane potential, as predicted by the Nernst equation.

$$\text{Nernst Potential for } Cl^- = 60/(-1) \log(100\,\text{mM}/10\,\text{mM}) = -60\,\text{mV}$$

Note that the Nernst potentials in examples 1 and 2 are both −60 mV, despite the opposite direction of concentration gradients—intracellular concentration is higher than extracellular concentration in example 1 but lower than extracellular concentration in example 2. This is because ionic valences in these two examples are opposite in sign—positive in example 1 but negative in example 2. Examples 1 and 2 illustrate the importance of considering concentration gradient and valence in the calculation of Nernst potential.

Example 3: Nernst Potential for Divalent Cation. Consider the following intracellular and extracellular concentrations of $CaCl_2$ for a cell, whose membrane is exclusively permeable to Ca^{2+}:

Intracellular fluid: 0.0001 mM $CaCl_2$

Extracellular fluid: 1 mM $CaCl_2$

The diffusion of Ca^{2+} down its concentration gradient, from extracellular fluid to the intracellular fluid, will result in a positive membrane potential, as predicted by the Nernst equation.

Table 3.1 Nernst Potentials of Na⁺, K⁺, and Cl⁻ in a Typical Mammalian Cell

	Na⁺	K⁺	Cl⁻
Extracellular Concentration	140 mM	4 mM	116 mM
Intracellular Concentration	12 mM	135 mM	4 mM
Nernst Potential	+64 mV	–92 mV	–88 mV

$$\text{The Nernst Potential for } Ca^{2+} = 60/(+2) \log (1 \text{ mM}/0.0001 \text{ mM}) = +120 \text{ mV}$$

As shown in examples 1 and 2, a tenfold concentration gradient for a monovalent cation will result in a Nernst potential of 60 mV. In comparison, as shown in example 3, a tenfold concentration gradient for a divalent ion will result in a Nernst potential of only 30 mV. At a given concentration gradient, the Nernst potential for divalent ion is lower than that for monovalent ion, because, in a given electric field, the driving force exerted on a divalent ion is twice as high as the driving force exerted on a monovalent ion.

IONIC BASIS OF RESTING MEMBRANE POTENTIAL

As shown in **Table 3.1**, the Nernst potentials for K⁺, Na⁺, and Cl⁻ for a typical mammalian cell are –92 mV, +64 mV, and –88 mV, respectively, as predicted by the concentration gradients for K⁺, Na⁺, and Cl⁻ across the cell membrane. As shown in Fig. 3.3, the resting membrane potential for a typical mammalian neuron is –70 mV, which is similar to the Nernst potentials for K⁺ and Cl⁻, but significantly different from the Nernst potential for Na⁺. As cited previously, ionic permeability is necessary for an ionic concentration gradient to establish a membrane potential. The resting membrane potential is similar to the Nernst potential for K⁺, because the cell membrane is most permeable to K⁺. The resting membrane potential is similar to the Nernst potential for Cl⁻, because the concentration gradient for Cl⁻ across the cell membrane is driven by the membrane potential.

Why is the Resting Membrane Potential Closest to the Nernst Potential for K⁺? It is important first to emphasize that Na⁺ – K⁺ – ATPase on the cell membrane establishes the concentration gradients of

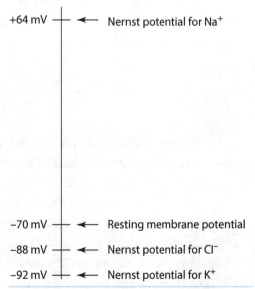

Fig. 3.3. Resting membrane potential in comparison with Nernst potentials for Na⁺, K⁺ and Cl⁻, as calculated from the extracellular concentrations shown in Table 3.1.

Na$^+$ and K$^+$ across the cell membrane, as shown in **Table 3.1**. The resting cell membrane is significantly more permeable to K$^+$ than Na$^+$ by approximately twentyfold. Consider the diffusion of K$^+$ and Na$^+$ across the cell membrane as driven by their concentration gradients before the establishment of resting membrane potential. The concentration gradient of K$^+$ (intracellular [K$^+$] > extracellular [K$^+$]) will drive the diffusion of K$^+$ out of the cell. In comparison, the concentration gradient of Na$^+$ (intracellular [Na$^+$] < extracellular [Na$^+$]) will drive the diffusion of Na$^+$ into the cell. The efflux of K$^+$ is larger than the influx of Na$^+$, because the cell membrane is more permeable to K$^+$ than Na$^+$. The net efflux of K$^+$ will cause the development of a negative cell membrane potential, which will drive the diffusion of both K$^+$ and Na$^+$ into the cell. The resting membrane potential cannot reach the Nernst potential for K$^+$, because the net flux of K$^+$ across the cell membrane will then be zero, whereas the net influx of Na$^+$ will be nonzero. Instead, the resting membrane potential will stabilize at a steady state that is slightly less negative than the Nernst potential for K$^+$, such that the net efflux of K$^+$ is exactly balanced by the net influx of Na$^+$.

Why is the Resting Membrane Potential Almost Identical to the Nernst Potential for Cl$^-$? In most cells, the concentration gradient of Cl$^-$ is not maintained by any ion-pumping ATPase, but mostly driven by the cell membrane potential established by K$^+$ and Na$^+$. The negative cell membrane potential drives the efflux of Cl$^-$, thereby establishing a concentration gradient of Cl$^-$ (extracellular [Cl$^-$] > intracellular [Cl$^-$]) for driving an equal influx of Cl$^-$. In red blood cells and skeletal muscle cells, the Nernst potential for Cl$^-$ is identical to the resting membrane potential. In some neurons, the Nernst potential for Cl$^-$ is slightly more negative than the resting membrane potential, because carrier-mediated Cl$^-$ transporters, in addition to the electrical gradient, regulate the concentration gradient of Cl$^-$ in these cells.

Goldman-Hodgkin-Katz (GHK) Equation. This equation predicts a membrane potential (V_m) as a function of multiple concentration gradients and membrane permeabilities, as shown in the following example for the resting membrane potential in a typical mammalian cell:

$$V_m = 60 \log \left(\frac{P_K[K]_0 + P_{Na}[Na]_0 + P_{Cl}[Cl]_i}{P_K[K]_i + P_{Na}[Na]_i + P_{Cl}[Cl]_0} \right)$$

P_K, P_{Na}, and P_{Cl} represent the permeabilities of K, Na, and Cl, respectively. The numerator includes extracellular concentrations of K$^+$ and Na$^+$ but intracellular concentration of Cl$^-$, because of the differences in valence—positive for K$^+$ and Na$^+$ and negative for Cl$^-$.

The GHK equation becomes the Nernst equation when the membrane is exclusively permeable to a single ion. For example, the GHK equation becomes the Nernst equation for K$^+$. when P_K is nonzero and P_{Na} and P_{Cl} are both zero. The GHK equation predicts that the ion having the highest permeability will dominate the numerator and denominator of the equation, causing the resting membrane potential to be close to the Nernst potential for the most permeable ion. This prediction implies that an increase in the cell membrane permeability to a specific ion will cause the cell membrane potential to shift toward the Nernst potential for that specific ion. For example, an increase in cell membrane permeability to Na$^+$ will cause the cell membrane potential to shift toward the Nernst potential for Na$^+$.

GRADED POTENTIALS AND SYNAPTIC POTENTIALS

Graded Potentials are local changes in cell membrane potential in response to stimuli in a strength-dependent manner. For example, a graded potential can be induced by the opening of ligand-gated ion channels, such that the amplitude of the graded potential is dependent on the ligand concentration. Generation of graded potentials by activating ligand-gated ion channels is a common mechanism by which neurotransmitters cause excitation or inhibition of a recipient cell. Graded potentials can be depolarization or hyperpolarization, depending on the type of ion channels involved. For example, graded depolarizations can be induced by the opening of ligand-gated Na^+ channels, and graded hyperpolarizations can be induced by the opening of ligand-gated K^+ channels. Graded potentials are generally not conducted over a long distance, but spatially confined within a short distance from the point of generation, because an ionic charge can leak through ion channels on the cell membrane during electrical conduction.

Ion Channels at Chemical Synapses. Chemical synapse is a signaling junction between two neurons or between a neuron and a muscle cell, at which a neurotransmitter released by the presynaptic neuron activates receptors on the postsynaptic cell. Fig. 3.4 shows the structure of a chemical synapse and mechanisms leading to the neurotransmitter release by the presynaptic neuron. As shown in Fig. 3.4, the presynaptic nerve terminal contains synaptic vesicles near the terminal, calcium channels on the cell membrane, and, depend-

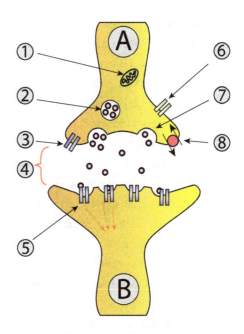

Fig. 3.4—Chemical Synapse. As shown in this figure, a presynaptic neuron (A) releases neurotransmitters into the synaptic cleft (4), where the neurotransmitter diffuses across the synaptic cleft to bind to receptors on the postsynaptic cell (B). A neurotransmitter may induce depolarization or hyperpolarization of the postsynaptic membrane depending on the type of receptor ion channels on the postsynaptic cell. Labels for this figure: 1) mitochondria; 2) synaptic vesicles for storing neurotransmitter; 3) auto-receptor ion channels on presynaptic neuron monitoring neurotransmitter release; 4) synaptic cleft; 5) receptor ion channels on postsynaptic membrane; 6) voltage-gated calcium channels for increasing intracellular $[Ca^{2+}]$ at the presynaptic terminal; 7) fusion of synaptic vesicles with presynaptic terminal membrane for neurotransmitter release; 8) re-uptake of neurotransmitter.

ing on the neuron type, may contain autoreceptors and enzymes for sensing and reuptake of the neurotransmitter, respectively. The postsynaptic cell membrane generally contains ligand-gated ion channels that are specific for the neurotransmitter released by the presynaptic neuron. The following mechanisms at the nerve terminal are typically involved in the stimulation of neurotransmitter release by presynaptic neurons: a) membrane depolarization; b) activation of voltage-gated calcium channels; c) increase in intracellular $[Ca^{2+}]$; d) Ca^{2+}-dependent fusion of synaptic vesicles with the terminal membrane; and e) exocytosis of neurotransmitter into the synaptic cleft.

Excitatory and Inhibitory Postsynaptic Potentials. A chemical synapse can be excitatory or inhibitory, depending on whether the neurotransmitter induces depolarization (excitation) or hyperpolarization (inhibition) of the postsynaptic cell. For example, the neuromuscular junction is an excitatory synapse, because the neurotransmitter released by the motor neuron causes depolarization of the muscle cell membrane. Depolarization of a postsynaptic cell in response to the neurotransmitter released from a presynaptic neuron is known as excitatory postsynaptic potential (**EPSP**). Conversely, hyperpolarization of a postsynaptic cell in response to the neurotransmitter released from a presynaptic neuron is known as inhibitory postsynaptic potential (**IPSP**).

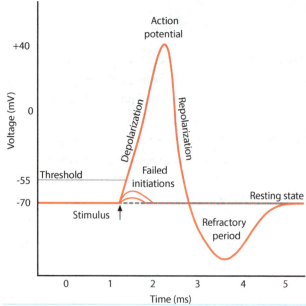

Fig. 3.5—Action Potential. Subthreshold depolarization fails to initiate an action potential, whereas suprathreshold depolarization triggers the generation of an action potential. The depolarization phase is caused by the opening of voltage-gated Na⁺ channels, and peaks near the Nernst potential for Na⁺. The repolarization phase is caused by the inactivation of Na⁺ channels and opening of voltage-gated K⁺ channels. The after-hyperpolarization phase is caused by the temporarily high K⁺ permeability of the cell membrane before the closing of all voltage-gated K⁺ channels. During absolute refractory period, it is impossible to trigger an action potential, because most Na⁺ channels remain inactivated. During relative refractory period, a stronger than normal stimulus is necessary for triggering an action potential, because some Na⁺ channels have come out of inactivation, but K⁺ permeability is relatively high.

ACTION POTENTIAL

Action Potential is a transient depolarization of the cell membrane that can be propagated to other regions of the cell—for example, from one end of an axon to the other end. Neurons use action potentials to send electrical signals over a long distance along axons. For example, a motor neuron in the spinal cord sends action potentials along an axon to stimulate neurotransmitter release at the neuromuscular junction that controls movement of a finger. Cardiac and skeletal muscle cells use action potentials to send electrical signals from the cell membrane toward the cell center to stimulate the release of Ca^{2+} from sarcoplasmic reticulum.

Fig. 3.5 shows the time course of an action potential in neurons—consisting of a depolarization phase, a repolarization phase, and an after-hyperpolarization phase. As shown in Fig. 3.5, an action potential is triggered by membrane depolarization that is above certain threshold potential. Ions are electric charge carriers in body fluids. Action potentials are caused by time-dependent changes in ionic movement across the cell membrane. The observation that a threshold depolarization is necessary for triggering an action potential suggests the involvement of voltage-gated ion channels

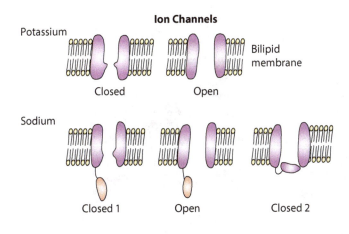

Fig. 3.6—Ionic Basis of an Action Potential. Voltage-dependent activation and inactivation of sodium channels and voltage-dependent activation of potassium channels are the basic mechanisms of action potential. Potassium channels open in response to depolarization and close in response to repolarization. Sodium channels open in response to depolarization and then become inactivated in the "closed 2" state with sustained depolarization. Repolarization is necessary for inactivated sodium channels to return to the "closed 1" state.

in the initiation of action potentials. A. L. Hodgkin and A. F. Huxley discovered the ionic mechanisms of action potentials as voltage and time-dependent regulation of Na^+ and K^+ permeability of the cell membrane.

Depolarization Phase of Action Potential. The **depolarization phase** of an action potential is caused by activation of voltage-gated Na^+ channels. As shown in Fig. 3.6 (**bottom panel; Na^+ channels**), when the cell membrane becomes depolarized above a threshold, voltage-gated Na^+ channels in the resting, closed state become activated (open). The resulting increase in Na^+ permeability causes the membrane potential to become depolarized further toward the Nernst potential for Na^+, until all voltage-gated Na^+ channels become activated and the membrane potential reaches peak depolarization. An action potential is said to be "all-or-none," because a supra-threshold depolarization will eventually cause the opening of all voltage-gated Na^+ channels, whereas a sub-threshold depolarization is insufficient to initiate an action potential.

Repolarization Phase of Action Potential. The **repolarization phase** of an action potential is caused by two mechanisms—depolarization-induced inactivation of Na^+ channels and activation of K^+ channel. With prolonged depolarization, Na^+ channels undergo inactivation to reach the inactivated, closed state—often depicted as occlusion of channel opening by an inactivation gate (Fig. 3.6, **bottom panel; Na^+ channels**). Na^+ channel inactivation decreases the Na^+ permeability of the cell membrane, thereby initiating the repolarization phase of an action potential. In addition, depolarization-induced activation of voltage-gated K^+ channels increases the K^+ permeability of the cell membrane, thereby continuing the repolarization phase of an action potential (Fig. 3.6, **top panel; K^+ channels**). Na^+ channels will remain in the inactivated, closed state, until repolarization of the cell membrane resets the inactivation gate, thereby changing the conformation of the Na^+ channel from the inactivated, closed state to the resting, closed state. In comparison, voltage-gated K^+ channels do not undergo inactivation after depolarization (Fig. 3.6, **top panel; K^+ channels**). The after-hyperpolarization phase of an action potential is caused by the slow return of K^+ channels from the open state to the resting state.

Absolute and Relative Refractory Periods. Absolute refractory period refers to the period of Na^+ channel inactivation immediately after the depolarization phase of an action potential. It is impossible to elicit action potentials during the absolute refractory period, because all Na^+ channels are in the inactivated, closed state. One function of the absolute refractory period is to ensure unidirectional propagation of action potentials along an axon. Relative refractory period refers to the period after absolute refractory period, but before the complete return to the resting membrane potential when some Na^+ channels have come out of the inactivated state and become available for activation. It is possible to elicit action potentials using strong stimuli during the relative refractory period (Fig. 3.5). The amplitude of action potentials generated during the relative refractory period tends to be smaller than normal, because some Na^+ channels remain inactivated and unavailable for action potential generation. During the relative refractory period, an extraordinarily strong stimulus is often necessary to induce depolarization of the cell membrane toward the threshold, because K^+ permeability of the cell membrane is higher than the resting level. Refractory periods sometimes function as a frequency filter in limiting the conduction of action potentials—for example, the conduction of impulses from pacemaker cells to the conductive system in the heart.

Propagation of Action Potentials. Action potentials can be propagated along an axon by repeatedly triggering action potentials. Depolarization phase of an action potential is conducted through intracellular and extracellular fluids to a neighboring region, thereby causing the region to become depolarized above the threshold for triggering an action potential. The process of depolarization and action potential generation is repeated along an axon until an action potential is generated at the end of an axon. The speed of propagation of action potentials by this process is limited by the distance of electrical conduction before depolarization falls below the threshold for eliciting an action potential, due to the loss of ionic charge through ion channels along the axon.

The speed of propagation of action potentials is faster in myelinated than unmyelinated axons,

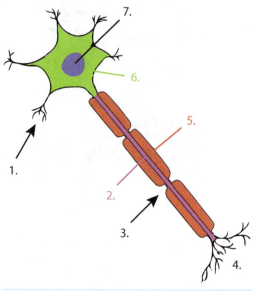

Fig. 3.7—Structure of a Myelinated Neuron. Dendrites (1) of a neuron receive excitatory and inhibitory inputs from other neurons via synapses. Neurotransmitters are synthesized in the cell body (6), as directed by the nucleus (7), packaged in synaptic vesicles, and transported to axon terminals (4) for release. The initial segment (2) is the site of action potential generation. Myelination (5) of an axon allows conduction of depolarization over a long distance for triggering action potentials at the unmyelinated Nodes of Ranvier (3).

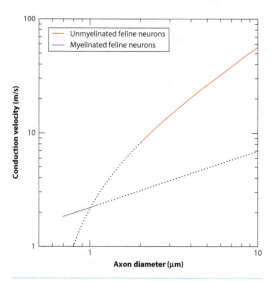

Fig. 3.8—Myelination Increases Velocity of Action Potential Conduction in Neurons. Unmyelinated neurons (blue) are relatively small in diameter and conduct action potentials at relatively slow velocity. In comparison, myelinated neurons (red) are relatively large in diameter and conduct action potentials at relatively fast velocity.

because myelination insulates parts of an axon from the loss of ionic charge through ion channels, thereby increasing the distance of electrical conduction before depolarization falls below the threshold for eliciting an action potential. As shown in Fig. 3.7, a myelinated neuron is made up of regions of myelination separated by nodes of Ranvier—the unmyelinated region for action potential generation. The "jumping" of action potentials at the nodes of Ranvier along a myelinated axon is known as **saltatory conduction.**

The speed of propagation of action potenials is faster in larger than smaller axons, because the intracellular electrical resistance of an axon is inversely proportional to the square of the axon diameter, as shown in Fig. 3.8. In general, large myelinated neurons have the highest conduction velocity of action potentials, whereas small unmyelinated neurons have the lowest conduction velocity.

Action potentials are used for cell–cell communications via chemical synapses. An action potential reaching the nerve terminal of a presynaptic cell induces the release of a neurotransmitter

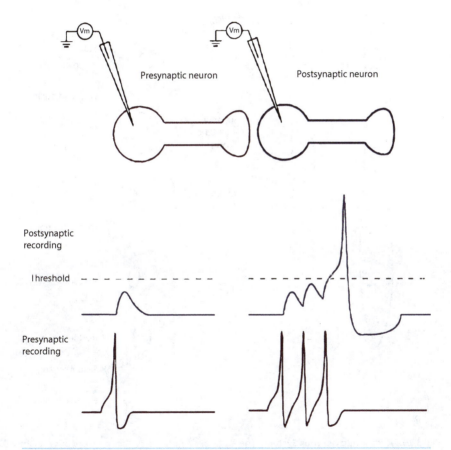

Fig. 3.9—Temporal Summation of Excitatory Post-synaptic Potentials (EPSPs). Top panel shows an excitatory synapse between two neurons. As shown in the left bottom panels, one action potential in the presynaptic neuron induces subthreshold depolarization of the postsynaptic neuron. As shown in the right bottom panels, temporal summation of a series of EPSPs triggered by a series of action potentials in the presynaptic neuron results in suprathreshold depolarization of the postsynaptic neuron for the generation of an action potential in the postsynaptic neuron.

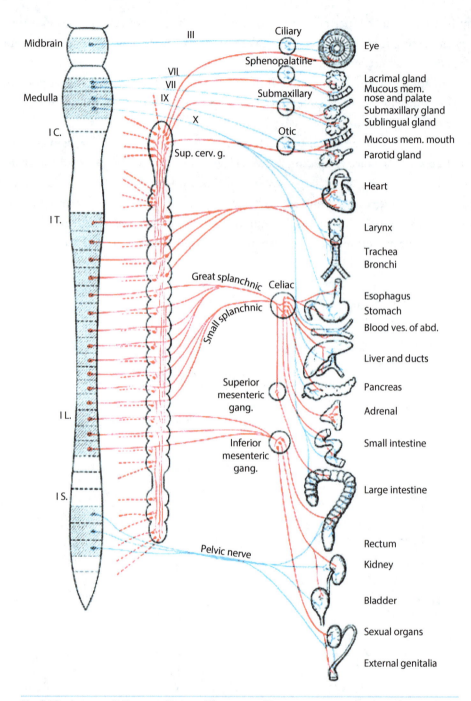

Fig. 3.10—Autonomic Nervous System. The autonomic nervous system consists of sympathetic nervous system (red) and parasympathetic nervous system (blue). The parasympathetic nervous system is dominant when the body is in a relatively relaxed state, whereas the sympathetic nervous system is dominant when the body is prepared for a fight or flight situation.

into the synaptic cleft, where the neurotransmitter diffuses to bind to and activate receptors on the postsynaptic cell, therefore causing either excitation (depolarization) or inhibition (hyperpolarization) of the postsynaptic cell.

CHEMICAL SYNAPSE AND ELECTRICAL SYNAPSE

In supra-threshold excitatory synapses—for example, neuromuscular junction—one action potential in a presynaptic cell is sufficient to induce depolarization in the postsynaptic cell that is above the threshold for eliciting an action potential. In sub-threshold excitatory synapses—for example, neuron–neuron synapses in the central nervous system—one action potential in a presynaptic cell induces depolarization of the postsynaptic cell that is below the threshold for eliciting an action potential in the postsynaptic cell, as shown in Fig. 3.9 (**left panel**).

Temporal Summation and Spatial Summation of Postsynaptic Potentials in Chemical Synapses. Temporal summation is the summation of postsynaptic potentials induced by a sequence of action potentials generated at close time intervals on the presynaptic cell. Fig. 3.9 (**right panel**) shows the temporal summation of sub-threshold excitatory postsynaptic potentials induced by a sequence of three action potentials generated at close time intervals, which results in a supra-threshold depolarization of the postsynaptic cell for eliciting an action potential in the postsynaptic cell.

Spatial summation is the summation of multiple postsynaptic potentials that are induced simultaneously by multiple presynaptic cells. Spatial summation can be additive when all synapses are uniformly excitatory or inhibitory, because ionic permeability changes induced by the synapses are additive in causing membrane depolarization or hyperpolarization. For example, when six excitatory presynaptic neurons form six synapses with the same postsynaptic neuron, activation of any one presynaptic neuron may induce an EPSP that is insufficient to elicit the generation of an action potential, whereas simultaneous activation of all six presynaptic neurons may be sufficient to induce an EPSP that is above the threshold for eliciting an action potential. Spatial summation can be subtractive when excitatory and inhibitory synapses are stimulated simultaneously, because ion permeability changes induced by excitatory and inhibitory synapses will counteract each other on their effect in membrane potential. For example, an excitatory presynaptic neuron normally induces an EPSP in the postsynaptic neuron by increasing Na^+ permeability of the postsynaptic neuron, whereas an inhibitory neuron normally induces an IPSP by increasing K^+ permeability of the postsynaptic cell. Simultaneous activation of the excitatory and inhibitory presynaptic neurons together will increase both Na^+ permeability and K^+ permeability of the postsynaptic neuron, resulting in a change in membrane potential that is intermediate between EPSP and IPSP induced by either presynaptic neuron alone.

Electrical Synapse is a path of direct electrical conduction between cells via gap junctions—nonspecific ion channels. By providing rapid passage of ions, gap junctions allow electrical synchronization between cells. For example, gap junctions between cardiac muscle cells enable cardiac muscle cells to contract synchronously for the ejection of blood into the circulation.

AUTONOMIC NERVOUS SYSTEM

Autonomic Nervous System is made up of two branches, the **parasympathetic nervous system** and the **sympathetic nervous system**. As shown in Fig. 3.10, most organ systems are regulated by both branches of the autonomic nervous system with opposite effects. The primary function of the parasympathetic nervous system is considered "house-keeping," because this system is most active when a person is in a relaxed state. The primary function of the sympathetic nervous system is considered "fight or flight," because this system is most active when a person is in an excited state.

Each branch of the autonomic nervous system consists of **pre-ganglionic and post-ganglionic fibers** that form synapses at the ganglia, where post-ganglionic fibers branch out to innervate target organs, as shown in the following scheme:

Pre-ganglionic Fibers → Ganglia → Post-ganglionic Fibers → Target Organs

As shown in Fig. 3.10, parasympathetic pre-ganglionic fibers originate from cranial nerves III, VII, IX, and X, and S2–S4 segments of the spinal cord. In comparison, sympathetic pre-ganglionic fibers originate from T1–T12, and L1–L3 segments of the spinal cord. Recognizing the origins of parasympathetic and sympathetic pre-ganglionic fibers from the central nervous system is important for understanding why injuries to different segments of the spinal cord will result in differential damages to the two branches of the autonomic nervous system. As shown in Fig. 3.10, parasympathetic ganglia are located mostly near organs, whereas sympathetic ganglia are located mostly near the spinal cord in the form of a ganglionic chain.

At both parasympathetic and sympathetic ganglia, pre-ganglionic fibers release the neurotransmitter **acetylcholine** (ACh), which binds to and **activates nicotinic acetylcholine receptors** (nAChRs) on post-ganglionic fibers, as shown in the following scheme:

Pre-ganglionic Fibers → ACh → nAChRs on Post-ganglionic Fibers

At target organs, parasympathetic and sympathetic post-ganglionic fibers release different neurotransmitters, which bind to and activate different receptors on cells. As shown in the following scheme, parasympathetic post-ganglionic fibers release acetylcholine (ACh), which binds and activates **muscarinic acetylcholine receptors** (mAChRs) on cells at target organs. In comparison, sympathetic post-ganglionic fibers release **norepinephrine** (NE), which binds to and activates **adrenergic receptors** on cells at target organs, as shown in the following scheme:

Parasympathetic Post-ganglionic Fibers → ACh → mAChRs at Target Organs

Sympathetic Post-ganglionic Fibers → NE → Adrenergic Receptors at Target Organs

In addition, sympathetic pre-ganglionic fibers innervate the adrenal medulla—central region of the adrenal gland. The adrenal medulla releases **epinephrine** (EPI) and norepinephrine (NE) into the circulation in response to sympathetic stimulation, as shown in the following scheme:

Sympathetic Pre-ganglionic Fibers → Adrenal Medulla → EPI and NE → Circulation

Epinephrine is similar to norepinephrine in structure. Both epinephrine and norepinephrine activate adrenergic receptors in target organs.

KEY TERMS

+ absolute refractory period
+ acetylcholine
+ action potential
+ adrenergic receptor
+ autonomic nervous system
+ chemical synapse
+ depolarization phase
+ electrical synapse
+ epinephrine
+ excitatory postsynaptic potential (EPSP)

+ Goldman-Hodgkin-Katz equation
+ graded potentials
+ inhibitory postsynaptic potential (IPSP)
+ muscarinic acetyl-choline receptor
+ Nernst potential
+ nicotinic acetylcholine receptor
+ norepinephrine
+ parasympathetic nervous system

+ post-ganglionic fiber
+ pre-ganglionic fiber
+ relative refractory period
+ repolarization phase
+ resting membrane potential
+ saltatory conduction
+ spatial summation
+ sympathetic nervous system
+ temporal summation

IMAGE CREDITS

4

MUSCLE PHYSIOLOGY

Muscle cells provide the shape and motion of many body organs, including the heart, arms, legs, airways, blood vessels, and uterus. There are three types of muscle cells. Skeletal muscle cells are attached to bones. The primary function of the musculoskeletal system is to maintain body posture and drive body movement. Cardiac muscle cells are found in the wall of the heart. The primary function of the cardiac muscle is to pump blood through the circulation. Smooth muscle cells are found in the wall of hollow organs—for example, airways, blood vessels, uterus, etc. The primary function of smooth muscle is to maintain the diameter of hollow organs against a transmural pressure.

Skeletal and cardiac muscle cells are structurally similar in being striated muscle cells, whose contractions are regulated by the thin filament-based troponin-tropomyosin system. Skeletal and cardiac muscle cells are significantly different in their dependencies on the nervous system for activation. As shown in Fig. 4.1, skeletal muscle contraction is absolutely dependent on activation by motor neurons projected from the central nervous system. Injuries to the central nervous system can result in skeletal muscle paralysis. In comparison, cardiac muscle contraction is activated by pacemaker

LEARNING OBJECTIVES

1. **Striated Muscle.** Discuss the structure and function of sarcomere in striated muscle cells.

2. **Neuromuscular Junction and Excitation-Contraction Coupling.** Discuss the structure and function of neuromuscular junction and explain how action potential generation on sarcolemma results in activation of skeletal muscle contraction.

3. **Crossbridge Cycle.** Describe the four phases of the crossbridge cycle, and discuss how cyclic interactions between myosin crossbridges and actin filament regulate crossbridge movement and myosin ATPase activity.

4. **Skeletal Muscle Contractions.** Compare and contrast isometric contraction and isotonic shortening in terms of defining characteristics and physiological function, and discuss the mechanisms underlying length–force and power–velocity relations.

5. **Motor Unit and Muscle Fiber Types.** Compare and contrast the three major skeletal muscle fiber types and motor units in terms of mechanical, metabolic, and motor characteristics.

6. **Muscle Reflexes.** Compare and contrast stretch reflex and Golgi tendon reflex in terms of structure and function.

7. **Cardiac Muscle and Smooth Muscle.** Compare and contrast skeletal muscle and cardiac muscle in terms of structure, electrophysiology, and intracellular $[Ca^{2+}]$ regulation; compare and contrast striated muscle and smooth muscle in terms of contractile filament structure and Ca^{2+}-dependent regulatory mechanism of contraction.

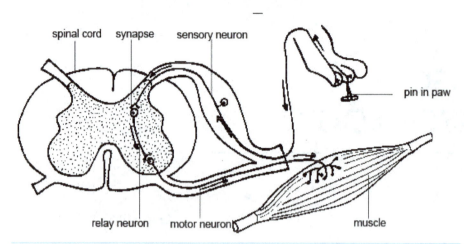

Fig. 4.1—Neural Control of Skeletal Muscle Contraction. This schematic diagram shows the spinal reflex for pain-induced paw retraction, which consists of sensory neurons that carry signals from pain receptors to the spinal cord, interneurons (relay neurons) that relay the sensory signal to the motor neurons, and motor neurons that stimulate skeletal muscle contraction.

cells situated within the heart, independent of the nervous system. The autonomic nervous system modulates heart rate and cardiac contractility, but is not necessary for eliciting spontaneous beating of the heart. Complete disconnection of a transplanted heart from the autonomic nervous system does not result in the cessation of beating of the heart, although the autonomic nervous system can no longer modulate cardiac muscle function directly. In comparison, smooth muscle cells are non-striated cells, whose contractions are regulated by the thick filament-based myosin light chain kinase system. These two regulatory systems will be examined in detail later in this chapter. Activation of smooth muscle cells can be induced by multiple means—for example, electrical stimulation, autonomic neurotransmitters, and local mediators.

STRUCTURE OF STRIATED MUSCLE CELLS

As shown in Fig. 4.2A, a whole skeletal muscle—for example, a bicep—consists of muscle bundles (fascicles), each of which holds individual muscle cells (fibers). As shown in Fig. 4.2B, microscopic examination of a skeletal muscle cell reveals striations, consisting of alternating dark (A) and light (I) bands. The A band contains thick filaments, of which the peripheral portion overlaps with the thin filaments and the central portion (H zone) does not overlap with the thin filaments. The I band—situated outside the A band—contains the central portion of thin filaments that does not overlap with myosin filaments. The Z line at the middle of the I band represents the anchoring of actin filaments by the actin-bundling protein α-actinin. As shown in Fig. 4.2B, a **sarcomere** is the basic contractile unit in a skeletal muscle cell, consisting of an A band and the two adjacent halves of I bands anchored by the two Z lines.

Fig. 4.3 illustrates the sliding filament model of muscle contraction, in which crossbridges on the thick filament pull on the overlapping thin filament, causing a sliding of actin filaments toward

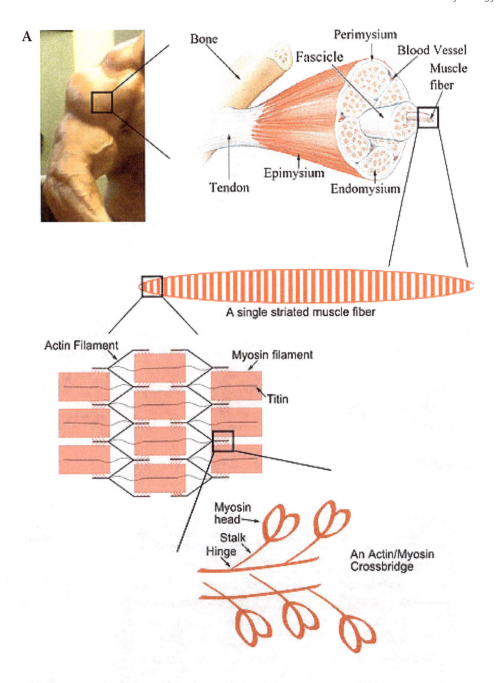

Fig. 4.2—Structure of Skeletal Muscle. Panel A shows a piece of skeletal muscle attached to a bone via tendon. Each skeletal muscle contains multiple muscle fascicles, each of which contains multiple skeletal muscle cells (fibers). When examined under a light microscope, a skeletal muscle fiber exhibits striations of light and dark bands. Panel B (next page) shows an electron microscopic image of a skeletal muscle cell and a schematic diagram of a sarcomere—basic contractile unit. The A band represents myosin filaments. The H zone in the middle of A band represents the central portion of myosin filaments that do not overlap with actin filaments. The I band represents thin filaments that do not overlap with myosin filaments. The Z line in the middle of I band represents actin-bundling proteins for the anchoring of actin filaments. The structure between two neighboring Z lines represents a sarcomere.

B

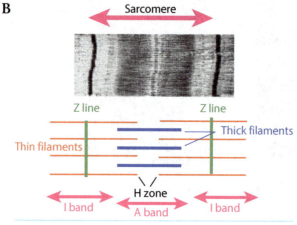

Fig. 4.2—Continued.

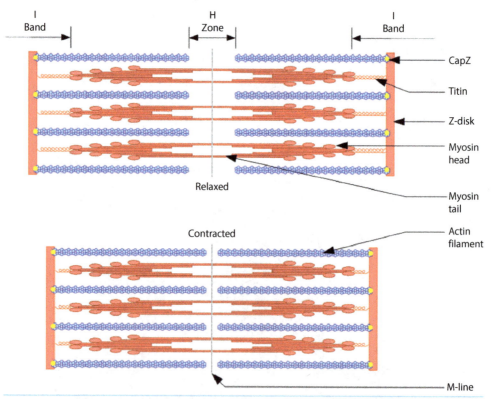

Fig. 4.3—Sliding of Actin Filaments During Muscle Shortening. The top panel shows a sarcomere in a relaxed muscle, where the ends of actin filaments from opposite sides are separated by a relatively wide H Zone. The bottom panel shows a shortened sarcomere in contracted muscle, where the ends of actin filaments from opposite sides have slid towards the center of the sarcomere, resulting in the narrowing of H Zone. CapZ is a protein that connects actin filaments to the Z line (Z-disk). Titin is a filament that connects myosin filaments to the Z-disk. M-line represents proteins for anchoring myosin filaments.

the center of the A band. As shown in Fig. 4.3, shortening of a striated muscle cell is accompanied by a narrowing of the I and H bands, when a large portion of the actin filaments overlaps with the myosin filaments in the A band. Muscle shortening has no effect on the width of the A band, which represents the entire length of myosin filaments.

NEUROMUSCULAR JUNCTION

Skeletal muscle contraction is dependent on activation by motor neurons projected from the central nervous system. Neuromuscular junction is the synapse between a skeletal muscle cell and the nerve terminal of a motor neuron. As shown in Fig. 4.4, an action potential arriving at the nerve terminal of a motor neuron causes fusion of synaptic vesicles with the terminal membrane and release of the neurotransmitter acetylcholine into the synaptic cleft. Acetylcholine then binds to nicotinic acetylcholine receptors on the muscle end plate, which causes depolarization of the muscle end plate toward the threshold for generation of an action potential on the muscle cell membrane. The neuromuscular junction is a suprathreshold synapse, because one action potential in the motor neuron is sufficient to elicit generation of action potential in the skeletal muscle end plate. Depolarization at the muscle end plate is then conducted to the neighboring cell membrane, causing the membrane potential to reach threshold for the generation of action potentials on the muscle cell membrane.

Fig. 4.5 shows the mechanism of neurotransmitter release by motor neurons at the

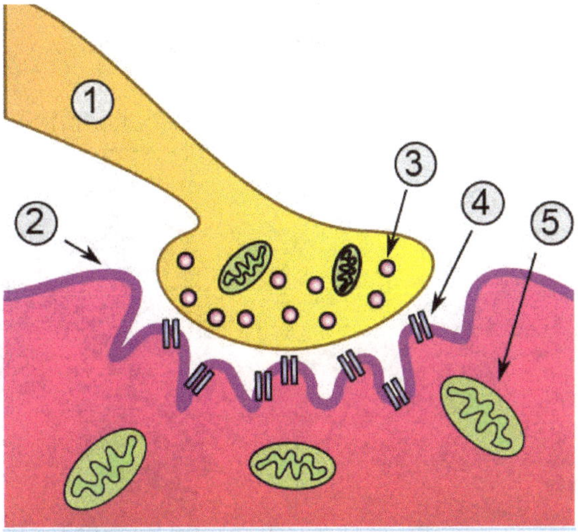

Fig. 4.4—Neuromuscular Junction—Synapse Between Motor Neuron (1) and Skeletal Muscle Cell Membrane (2). An action potential arriving at the nerve terminal of a motor neuron causes the fusion of synaptic vesicles (3) with the nerve terminal and release of acetylcholine into the synaptic cleft. Acetylcholine then binds to the nicotinic acetylcholine receptors (4) on the skeletal muscle cell membrane, causing depolarization of the skeletal muscle cells membrane and generation of an action potential. Mitochondria (5) are present in skeletal muscle cells for ATP production.

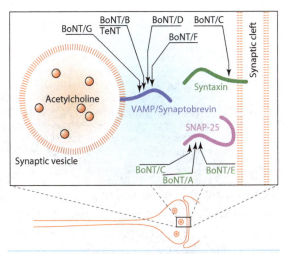

Fig. 4.5—Botulinum toxin (BoNT) induces skeletal muscle relaxation by inhibiting proteins (syntaxin, SNAP-25, VAMP/Synaptobrevin) for the docking of synaptic vesicles at the terminal of a motor neuron.

neuromuscular junction, which is the scientific basis of **Botox**-induced skeletal muscle relaxation in

Skeletal Muscle Fiber

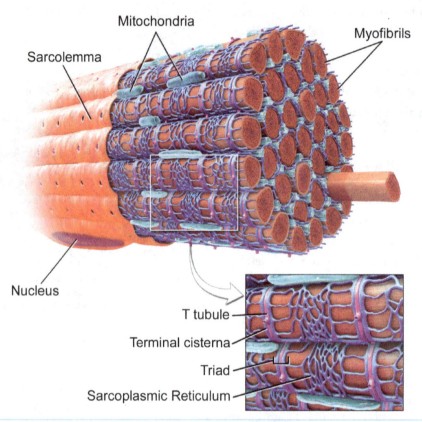

Fig. 4.6—Excitation-contraction Coupling in Skeletal Muscle. This figure shows the relative locations of transverse tubules and sarcomeres in a skeletal cell. Small panel shows the extension of transverse tubules from the plasma membrane into the center of a skeletal cell. Triad is the interface between a transverse tubule and the sarcoplasmic reticulum via the calcium release units, consisting of dihydropyridine receptors in the T tubule and ryanodine receptors. Dihydropyridine receptors function as a voltage sensor. Ryanodine receptors are calcium channels for the release of Ca^{2+} from sarcoplasmic reticulum. SERCA is a Ca^{2+}-ATPase in the sarcoplasmic reticulum membrane for pumping Ca^{2+} from the cytoplasm to the sarcoplasmic reticulum.

the treatment of wrinkles and migraine headache. As shown in Fig. 4.5, docking of a synaptic vesicle with the terminal membrane of a motor neuron is mediated by the interaction between a synaptic vesicle-associated protein (synaptobrevin) and two nerve terminal–associated proteins (SNAP-25 and syntaxin). Botulinum toxins (Botox) are capable of cleaving these three synaptic proteins, thereby inhibiting the release of acetylcholine by motor neurons at the neuromuscular junction, and causing relaxation of skeletal muscle cells for the treatment of wrinkles and migraine headache.

EXCITATION-CONTRACTION COUPLING

"Excitation-contraction coupling" refers to the mechanism by which membrane depolarization leads to muscle contraction. As covered previously, the release of the neurotransmitter acetylcholine at the neuromuscular junction results in membrane depolarization at the muscle end plate, which then causes depolarization of the muscle cell membrane. As shown in Fig. 4.6, action potentials generated on the muscle cell membrane propagate through transverse tubules (T tubules) to reach the middle of a muscle cell, where dihydropyridine receptors on the transverse tubule make contact with ryanodine receptor calcium channels on the sarcoplasmic reticulum. Depolarization-induced activation of dihydropyridine receptors on the transverse tubule is coupled to the activation of ryanodine receptor calcium channels on the sarcoplasmic reticulum, causing the release Ca^{2+} from the sarcoplasmic reticulum into the cytoplasm. As shown in Fig. 4.7, Ca^{2+}-binding to **troponin** causes conformational change of troponin, which in turn causes conformational

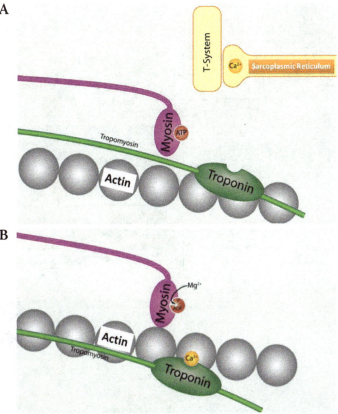

Fig. 4.7—Ca^{2+}-Troponin-Tropomyosin System for the Regulation of Muscle Contraction. Panel A shows the inactive system in a relaxed muscle, where intracellular Ca^{2+} concentration is low, because Ca^{2+} is sequestered in the sarcoplasmic reticulum. In the absence of Ca^{2+} binding to troponin, tropomyosin blocks crossbridge binding sites on actin filament and prevents the interaction between myosin crossbridges and actin filament for contraction. Panel B shows the active system in a contracted muscle, intracellular Ca^{2+} concentration is high, because Ca^{2+} is released from the sarcoplasmic reticulum to the cytoplasm. Binding of Ca^{2+} to troponin induces conformational change of the troponin-tropomyosin complex, thereby removing the blocking effect of tropomyosin on crossbridge binding sites on actin filament. Myosin crossbridges are free to interact with actin filament for muscle contraction.

change of **tropomyosin**, resulting in the availability of myosin-binding sites on the actin filament for binding by myosin crossbridges. Muscle force and shortening are generated by cyclic interactions between actin filament and myosin crossbridges.

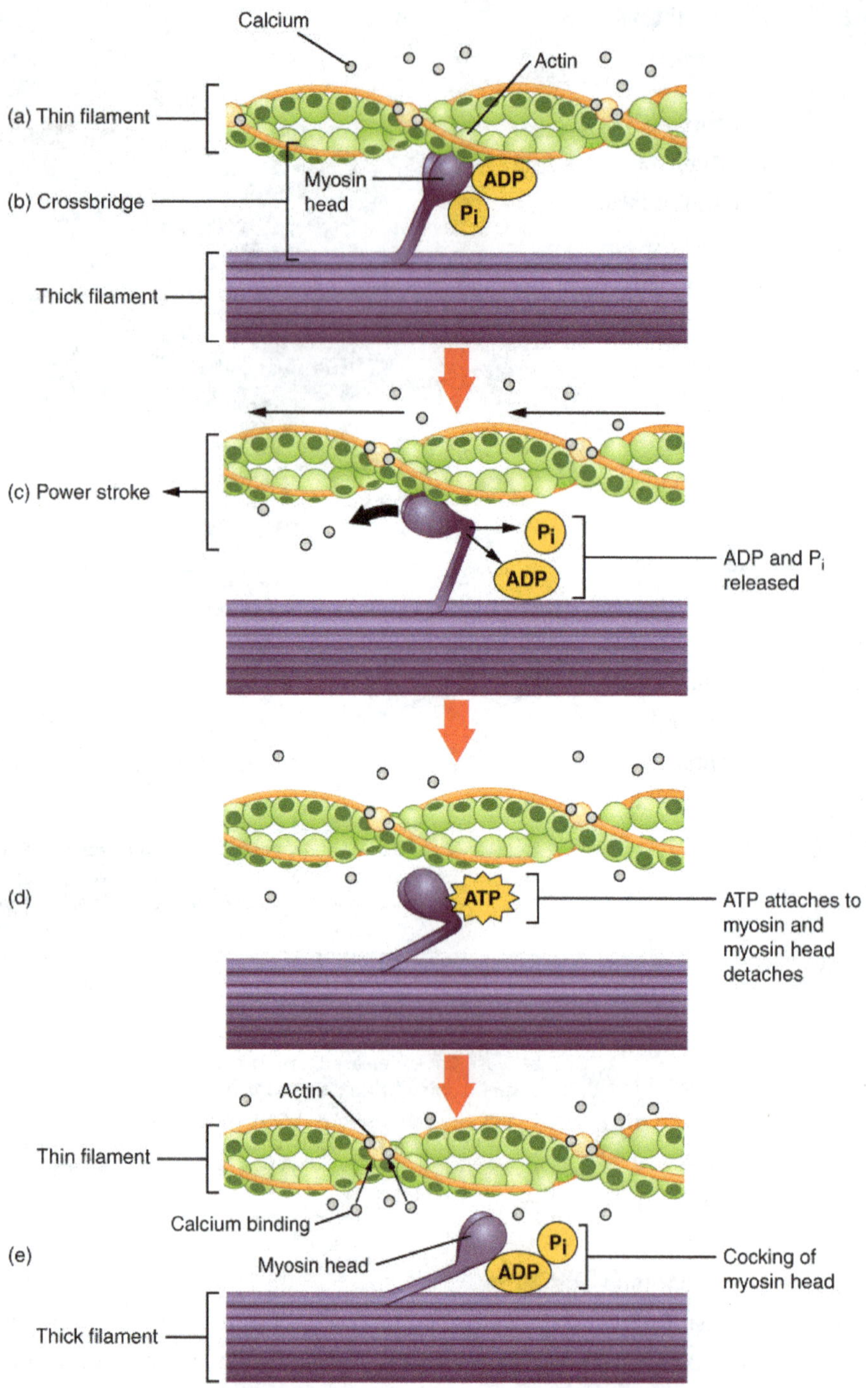

Fig. 4.8—Crossbridge Cycle. This figure shows the cyclic interactions between myosin cross-bridges and the actin filament during muscle contraction. (d) Detached crossbridge after binding of ATP to the crossbridge at the end of a powerstroke when the globular head of myosin is remains at a 45° angle; (e) Detached, energized crossbridge after partial hydrolysis of ATP to ADP.P_i and resetting of the globular head of myosin to 90° angle (b) Attached crossbridge ready to undergo a powerstroke; (c) Attached crossbridge after a powerstroke, complete hydrolysis of ATP, and release of ADP and P_i. During a power stroke, the globular head of a myosin crossbridge rotates from 90° to 45° angle, thereby exerting pulling force on the actin filament.

CROSSBRIDGE CYCLE

Crossbridge Cycle refers to the cyclic interactions between actin filament and myosin crossbridges, during which chemical energy released by ATP hydrolysis is coupled to the movement of myosin crossbridges. The coupling between myosin-ATPase activity and mechanical movement of myosin crossbridge is mediated by interactions between actin and ATP-binding sites on myosin crossbridges. Binding of actin to myosin is necessary for myosin-ATPase activity, which is analogous to the accelerator pedal for driving a car. Myosin-ATPase activity is low in a relaxed muscle, because actin is unavailable for binding by myosin. In a contracting muscle, when actin is available for binding by myosin, the cyclic interactions between actin and myosin crossbridge results in ATP hydrolysis with the release of energy that is transduced to crossbridge movement.

Fig. 4.8 shows the major steps during a crossbridge cycle, in which cyclic attachment and detachment of myosin crossbridges to and from the actin filament are coupled with ATP hydrolysis and angular movement of myosin crossbridge. Panel d in Fig. 4.8 is a good starting point for discussing the crossbridge cycle. Fig. 4.8d shows a detached ATP-bound myosin crossbridge with a tilted (45°) angle that has just detached from the actin filament. Fig. 4.8e shows the crossbridge in an "energized" state, when ATP is partially hydrolyzed to ADP.Pi by myosin ATPase activity and myosin crossbridge is reset to a vertical (90°) angle. Myosin crossbridges are maintained in the energized state in a relaxed muscle cell, because complete hydrolysis of ADP.Pi cannot proceed when actin is not available for binding by myosin. In an activated muscle cell where actin is available for binding, an energized myosin crossbridge becomes attached to the actin filament first in the weakly bound state (Fig. 4.8b), and subsequently undergoes a conformational change ("power stroke") to the strongly bound state with a tilted (45°) angle after complete hydrolysis of ATP to ADP and Pi and release of the products from myosin (Fig. 4.8c). Power stroke of an attached myosin crossbridge is the molecular mechanism by which a myosin crossbridge propels the sliding of an actin filament. As shown in Fig. 4.8d, binding of ATP to an attached crossbridge in the strongly bound state causes detachment of the crossbridge from the actin filament.

Shortening of a contracting muscle is driven by continuous but asynchronous attachment and detachment of myosin crossbridges to and from the actin filament. Asynchronous cycling of crossbridges is essential for force maintenance, because some crossbridges must remain attached to the actin filament for producing force at any given time. If cycling of all crossbridges were synchronized, then force would fall to zero when all crossbridges simultaneously detach from the actin filament. The continuous attachment and detachment of crossbridges during muscle contraction require a stable intracellular [ATP] that is maintained by energy metabolism. Skeletal muscle becomes stiff—a condition known as rigor mortis—sometime after death, when intracellular [ATP] becomes depleted, because crossbridges cannot detach from the actin filament in the absence of ATP, as predicted by the crossbridge cycle.

A

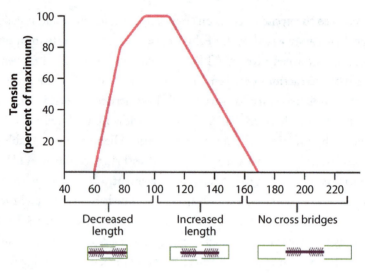

Percentage sarcomere length

B

Percentage sarcomere length

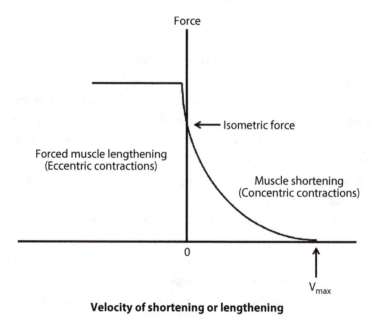

Velocity of shortening or lengthening

Fig. 4.9—A. Isometric sarcomere length-force relation—consisting of ascending limb, plateau, and descending limb. Optimal length is the sarcomere length at which active force reaches maximum.   **B.** Isotonic force-velocity relation for concentric contractions (muscle shortening) and eccentric contractions (forced muscle lengthening).

MUSCLE MECHANICS

Force production and muscle shortening are the two main functions of skeletal muscle cells. For example, maintenance of a posture and weight-holding require force production from skeletal muscle contraction. In comparison, walking and running require both force production and muscle shortening by skeletal muscles. Two types of experiments—isometric contraction and isotonic shortening—have been designed to study the regulation of force production and muscle shortening by skeletal muscles.

Isometric Contraction. Isometric contraction is force generation at constant muscle length. Contraction of core muscles for stabilization of the spinal column is an example of isometric contraction. Isometric contraction experiments are designed to study force development by a muscle as a function of muscle length. In an isometric experiment, one end of the muscle is attached to a force transducer for force measurement, and the other end of the muscle is attached to a length manipulator for setting muscle length. During an experiment, passive and active forces developed by a muscle are measured at various muscle lengths. In a relaxed muscle, passive force is developed by the stretching of elastic filaments inside and outside a muscle cell. As shown in Fig. 4.9A, active force developed by a contracting muscle is length-dependent. Active force begins to develop when sarcomere length is longer than 1.2 μm, and then increases with sarcomere length up to a maximum at 2 μm. The sarcomere length or muscle length at which active force reaches maximum is defined as optimal length. Increase in sarcomere length beyond optimal length results in the decline in active force. At 3.6 μm sarcomere length, a skeletal muscle cell can no longer develop any active force.

The characteristic length-active force relation in skeletal muscle (Fig. 4.9A) can be explained by the sliding filament hypothesis in terms of length-dependent changes in the overlap between actin and myosin filaments. When sarcomere length is shorter than optimal length, active force is submaximal, because actin filaments situated on either side of a thick filament slide beyond the center of the thick filament, causing interference with each other in the interaction with myosin crossbridges. Increase in sarcomere length up to optimal length gradually removes the interference between actin filaments situated on either side of a thick filament, thereby increasing active force as a function of sarcomere length. Active force reaches a maximum at optimal length, because actin filaments from either side of a thick filament overlap maximally with crossbridges on the thick filament. When sarcomere length is longer than optimal length, the central portion of a thick filament no longer overlaps with actin filaments, causing some myosin crossbridges to lose access to actin filament for attachment and active force development. At extremely long sarcomere lengths (> 3.6 μm) active force falls to zero, because actin filaments become completely stretched beyond myosin filaments, and all myosin crossbridges lose access to actin filament for attachment and active force development. It is noteworthy that length-dependent muscle activation, as measured by intracellular $[Ca^{2+}]$, also plays a significant role in the regulation of active force production as a function of muscle length.

Isotonic Shortening. Isotonic shortening is muscle shortening against constant force. Lifting of a weight is an example of isotonic shortening. In an isotonic shortening experiment, one end of a muscle is secured to a fixed point, and the other end of the muscle is attached to a loading device that can be programmed to produce a specific external load to the muscle. Fig. 4.9B shows the force-velocity relationship in a contracting muscle. The abscissa in Fig. 4.9B includes both negative and positive shortening velocities.

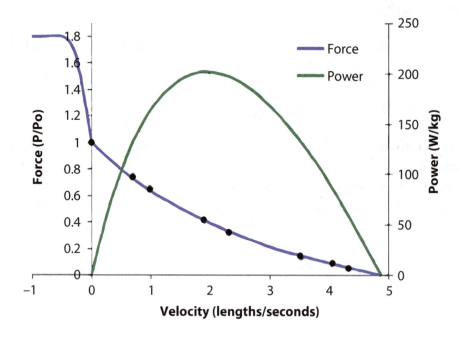

Fig. 4.10—Mechanical Power Output from Skeletal Muscle as a Function of Shortening Velocity. Mechanical power output reaches maximum at intermediate shortening velocity.

Positive shortening velocities represent **concentric** contractions, when a muscle shortens against external loads that are lower than the maximum force (isometric force) that can be developed by the muscle. Lifting a weight is an example of concentric contraction. As shown in Fig. 4.9B, during concentric contractions, the force-velocity relationship is characterized by a hyperbolic curve. Shortening velocity is zero when external load exactly matches the maximum force that can be developed by the muscle. By definition, zero shortening is identical to isometric contraction. Shortening velocity is positive when the external load is lower than the maximum force that can be developed by a muscle. Shortening velocity reaches a maximum level (V_{max}) when external load is zero.

Negative shortening velocities represent **eccentric** contractions, when a muscle is forced to lengthen by external loads that are higher than the maximum force (isometric force) that can be developed by the muscle. As shown in Fig. 4.9B, during eccentric contractions, yielding force developed by a muscle during forced lengthening is approximately 150% isometric force. Eccentric contractions are necessary for forced extension of a calf muscle during downhill walking and running. Repeated eccentric contractions can lead to muscle damage and soreness.

Mechanical Power Output from Muscle Contraction. Mechanical power equals shortening velocity × force. Mechanical power output is zero in both isometric contraction and unloaded shortening, because shortening velocity is zero in isometric contraction and external force is zero in unloaded shortening. It is noteworthy that, despite the lack of mechanical power output from muscle, continuous ATP consumption is necessary for supporting the continuous crossbridge cycling that both isometric contraction

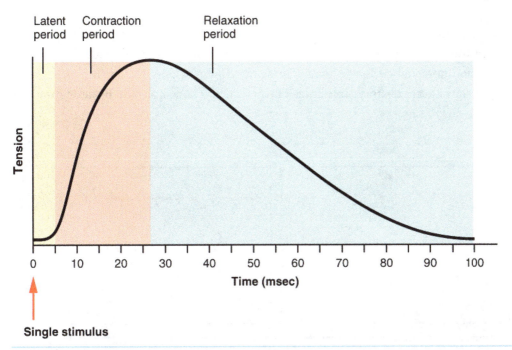

Fig. 4.11—Muscle witch triggered by one action potential.

and unloaded shortening require. Mechanical power output by muscle is nonzero during body movements—for example, running and bicycling, because both shortening velocity and force are nonzero. As shown in Fig. 4.10, mechanical power output by muscle varies as a function of shortening velocity and force. Mechanical power output is maximal at intermediate shortening velocity and intermediate external force—approximately 30% of maximal force. Gears in a bicycle are designed to allow a rider to maximize mechanical power output from muscle for moving the bicycle at the highest speed.

MUSCLE TWITCH, TWITCH SUMMATION AND TETANUS

Muscle twitch is the transient contraction of skeletal muscles in response to one action potential. Force development during a muscle twitch is transient because intracellular $[Ca^{2+}]$ induced by one action potential is transient. Ca^{2+} is first released from sarcoplasmic reticulum in response to the depolarization phase of an action potential. Ca^{2+} is then taken up by the sarcoplasmic reticulum in response to the repolarization phase of an action potential. As shown in Figure 4.11, a muscle twitch consists of three phases. The latent period between application of action potential and onset of contraction represents the time necessary for intracellular Ca^{2+} release, crossbridge activation, and crossbridge cycling. The contraction period represents an increase in force (tension) as a result of crossbridge attachment induced by the rise of intracellular $[Ca^{2+}]$. The relaxation period represents a decrease in force as a result of crossbridge detachment induced by the fall of intracellular $[Ca^{2+}]$ back to basal level.

It is noteworthy that the duration of an action potential in a skeletal muscle cell is typically 1 msec, whereas the contraction period is typically 20 msec. It is therefore possible to stimulate a skeletal muscle cell repetitively with a train of multiple action potentials during the contraction period. As shown in Fig. 4.12A, when the frequency of the train of action potentials is low, that is, when the time separation between action potentials is longer than contraction time of a muscle twitch, muscle twitches partially overlap, resulting in oscillatory force development, which is known as **twitch summation**. Force

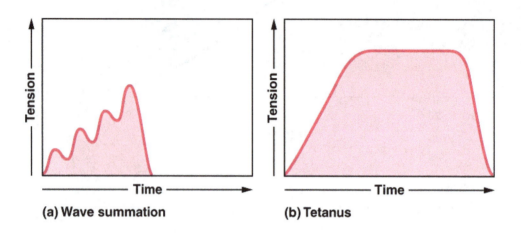

(a) Wave summation   **(b) Tetanus**

Fig. 4.12—A. Wave summation—partial overlapping of multiple muscle twitches triggered by a train of action potentials at low frequency. **B.** Tetanus—complete overlapping of multiple muscle twitches triggered by a train of action potentials at high frequency.

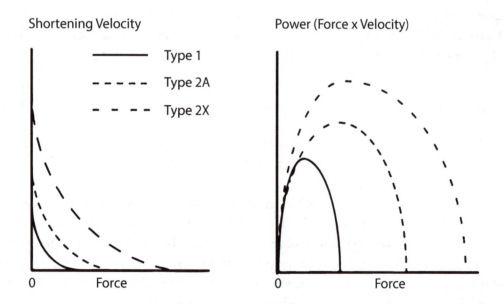

Fig. 4.13—Force-velocity and Force-power Relations for Slow and Fast Muscle Fibers. Type 1 slow muscle fibers exhibit low unloaded shortening velocity and low mechanical power output, but are relatively fatigue-resistant. Type 2 Fast muscle fibers exhibit relatively high unloaded shortening velocity and high mechanical power output, but are prone to muscle fatigue.

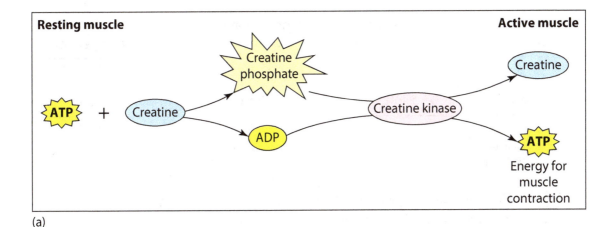

(a)

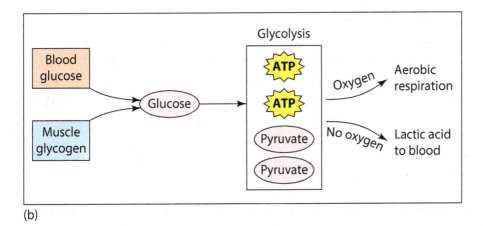

(b)

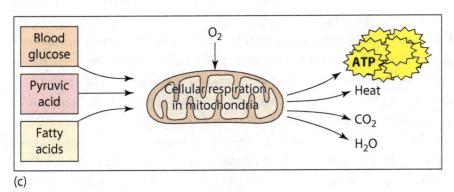

(c)

Fig. 4.14—Anaerobic Glycolytic and Aerobic Mitochondrial Metabolic Pathways. Anaerobic glycolysis is fast but non-sustainable, whereas oxidative phosphorylation is slow but sustainable. Fast muscle fibers utilize anaerobic glycolysis for energy. Slow muscle fibers utilize aerobic mitochondrial metabolic pathways for energy.

Motor unit type	Frequency of action potentials	Firing duration per train	Number of action potentials per day
Slow	Low	Long	High
Fast	High	Short	Moderate

Fig. 4.15. Different patterns of action potential firing in low and fast motor units.

developed during twitch summation is higher than force developed during a single muscle twitch due to the buildup of intracellular $[Ca^{2+}]$ as a result of repetitive release of Ca^{2+} from the sarcoplasmic reticulum before complete reuptake of Ca^{2+} by the sarcoplasmic reticulum. As shown in Fig. 4.12B, when the frequency of the train of action potentials is high, that is, when the time separation between consecutive action potentials is shorter than contraction time of a muscle twitch, muscle twitches completely overlap, resulting in smooth development of force without oscillations, which is known as **tetanus.** Force developed during tetanus is higher than force developed during a single muscle twitch or twitch summation due to the higher intracellular $[Ca^{2+}]$ in tetanus as a result of repetitive release of Ca^{2+} from the sarcoplasmic reticulum without intermittent relaxation.

MOTOR UNITS AND MUSCLE FIBER TYPES

Muscle cells are organized in groups—motor units—for sequential recruitment in response to the demand of a given task. A motor unit consists of a motor neuron and the muscle fibers it innervates. Within each motor unit, the firing frequency of action potentials of the motor neuron is generally matched to the speed of the muscle fiber type. For example, a small motor unit is made up of a low-frequency motor neuron and a small number of slow muscle fibers. Small motor units are recruited for fine movements—for example, walking, writing, and chewing. In comparison, a large motor unit consists of a high-frequency motor neuron and a large number of fast muscle fibers. Large motor units are recruited for fast and powerful movements—for example, jumping.

Muscle Fiber Types.   There are three major muscle fibers—slow oxidative (type 1), fast oxidative glycolytic (type 2A), and fast glycolytic (type 2X) fibers. As shown in Fig. 4.13, different muscle fiber types exhibit distinct force-velocity and force-power relations. For example, slow oxidative (type 1) muscle fibers are only capable of generating low unloaded shortening velocity and low maximum mechanical power output. Type 1 fibers are relatively fatigue-resistant and utilized for sustained fine movements—for example, writing and walking. In comparison, fast glycolytic (type 2X) muscle fibers are capable of generating high unloaded shortening velocity and high maximum mechanical power output. Type 2X fibers are fatigue-prone and utilized for short-term, fast, and powerful movements—for example, jumping.

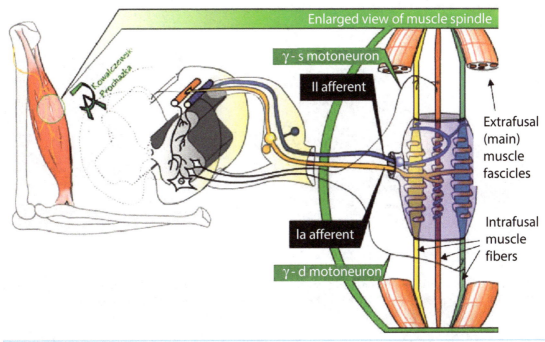

Fig. 4.16—Intrafusal Fibers Function as Length Sensor in the Stretch Reflex. Afferent sensory neurons (Ia and II) carry stretch signals to the spinal cord. γ-Motor neurons (γ and γ-d) control the length of intrafusal fibers to match the length of extrafusal muscle fibers.

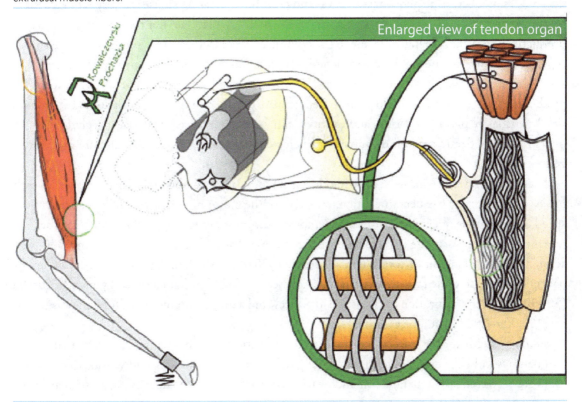

Fig. 4.17. Golgi tendon organ functions in an inhibitory reflex for preventing the detachment of muscle from tendon.

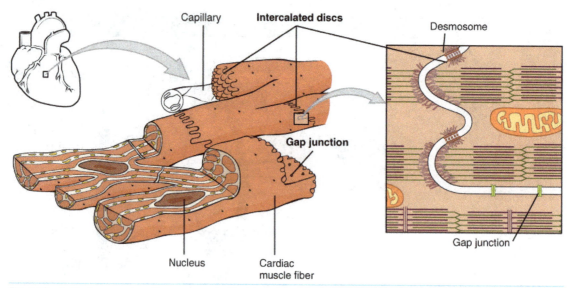

Fig. 4.18—A. Cardiac muscle cells are connected by intercalated discs.   **B.** Expanded diagram showing that an intercalated disc contains desmosome for physical connection, and gap junctions for electrical connection.

Different muscle fiber types utilize different metabolic pathways for ATP production. As shown in Fig. 4.14A, Creatine phosphate functions as a buffer for maintaining intracellular [ATP], because the conversion of creatine phosphate to ATP and creatine by creatine kinase is a very rapid reaction. As shown in Fig. 4.14B, glycolysis is a fast metabolic pathway that does not require oxygen for ATP production, but is non-sustainable due to the accumulation of lactate as a byproduct. Fast glycolytic type 2X fibers are fatigue-prone due to lactate accumulation from anaerobic glycolysis. As shown in Fig. 4.14C, oxidative phosphorylation is a slow metabolic pathway that requires oxygen for ATP production, but is sustainable. Slow oxidative fibers are enriched in mitochondria for ATP production by oxidative phosphorylation, and relatively fatigue-resistant. Fast oxidative glycolytic type 2A fibers have a moderate amount of mitochondria for oxidative ATP production and are moderately fatigue-resistant.

Matching Between Motor Neuron and Muscle Fiber Type. As shown in Fig. 4.15, within each motor unit, the firing frequency of action potentials of the motor neuron is generally matched to the speed of the muscle fiber type. For example, motor neurons innervating the slow oxidative fiber, soleus muscle, are characterized by a low frequency of firing (15–20/sec), long-lasting trains (300–600 secs), and a high amount of total impulse activity (300,000–500,000/day). In comparison, motor neurons innervating the fast oxidative glycolytic fiber, EDL-2, are characterized by high frequency (50–90/sec), a short duration of trains (60–150 sec), and a moderate amount of total impulse activity (100,000–250,000 impulses/day).

Sequential Motor Unit Recruitment. In daily activities, slow motor units are recruited first to handle slow and light motor activities with long duration such as walking and writing. Fast motor units are recruited last to handle fast and heavy motor activities with short duration such as lifting a heavy object.

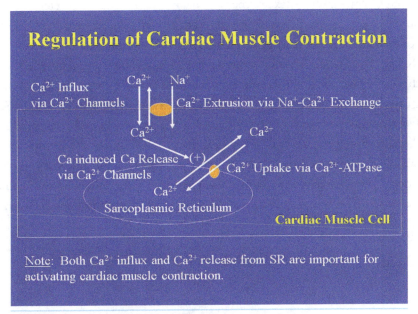

Fig. 4.19. Intracellular Ca²⁺ regulation in cardiac muscle cell.

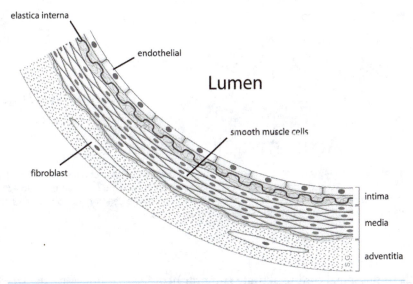

Fig. 4.20. Section of a carotid artery showing the middle layer of vascular smooth muscle cells in the media, single luminal layer of endothelial cells in the intima, and outer layer of connective tissue in the adventitia.

MUSCLE REFLEXES

Stretch reflex and Golgi tendon reflex are the two major muscle reflexes. Stretch reflex functions in stabilizing muscle length against external perturbations to a muscle, whereas the Golgi tendon reflex functions in preventing tendon breakage by excessive muscle contraction.

Stretch Reflex. The neural circuit for stretch reflex consists of intrafusal muscle fibers for sensing muscle length, sensory neurons for carrying stretch signals to the spinal cord, and α-motor neurons for activating extrafusal muscle fibers. Fig. 4.16 shows two types of intrafusal muscle fibers, two types of sensory neurons, and two types of motor neurons in a muscle. Ia and II afferent neurons carry dynamic and static stretch signals from intrafusal fibers to the spinal cord, respectively. Dynamic (γ-d) and static (γ-s) motor neurons control the lengths of dynamic and static intrafusal muscle fibers, respectively. Stretch reflex maintains muscle length by the following mechanisms. Sudden stretching of a muscle by external force causes the stretching of intrafusal muscle fibers, which send

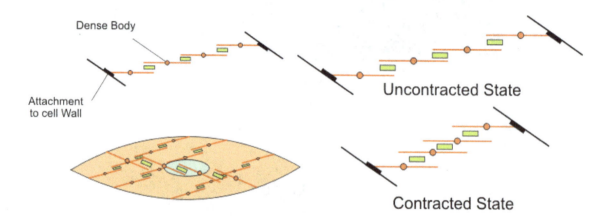

Dense Body

Attachment to cell Wall

Uncontracted State

Contracted State

Actin-myosin filaments

Fig. 4.21—Structure of a Smooth Muscle Cell. The basic contractile unit consists of actin filaments (brown lines) and myosin filaments (green rectangle). Actin filaments are anchored to dense plaques in the cell membrane (black rectangles) and dense bodies in the cytoplasm (brown circles). Shortening of a smooth muscle cell from uncontracted state to contracted state is mediated by the sliding of actin filaments by myosin crossbridges.

sensory signals to the spinal cord to activate α-motor neurons that innervate the stretched muscle, resulting in an increase in muscle contraction and restoration of muscle length. Stimulation of the knee-jerk stretch reflex is often used clinically to test the integrity of the neural circuitry. When the stretch reflex is functional, tapping the knee tendon stimulates contraction of the muscle connected to the tendon, resulting in extension of the lower leg. During normal body movement, the stretch reflex is active and functional in maintaining posture against sudden perturbations. For example, an unanticipated stepping of one foot into a hole does not cause a person to fall into the hole, because the loss of stretch on the falling leg will inhibit further stepping of the foot from the hole. In addition, temporary leaning of the body toward the hole as a result of stepping into the hole will stimulate the stretch reflex of the body muscle to restore the body posture.

Coactivation of α and γ-Motor Neurons in Motor Unit Recruitment. Alpha-motor neurons control the activation of extrafusal muscle fibers that power body movement. In comparison, as shown in Fig. 4.16, γ-motor neurons control the contraction of intrafusal muscle fibers, which is necessary for maintaining tautness of intrafusal muscle fibers for sensing changes in muscle length. During motor unit recruitment, the central nervous system simultaneously activates α-motor neurons to extrafusal fibers and γ-motor neurons to intrafusal fibers—a process known as **alpha-gamma coactivation**—to ensure that intrafusal muscle fiber length matches extrafusal fiber length.

The Golgi Tendon Reflex is an inhibitory reflex for preventing the rupture of a tendon by excessive muscle contraction. As shown in Fig. 4.17, excessive stretch of the tendon is sensed by the Golgi tendon apparatus, which sends signals through afferent neurons to the spinal cord to inhibit muscle contraction. Despite the presence of the Golgi tendon reflex, rupture of the Achilles tendon can still occur occasionally when excessive force is generated rapidly by a strong contracting muscle.

Reciprocal Inhibition in Locomotion. Joint movement is controlled by antagonistic muscle groups. For example, bending of a knee is driven by flexor muscles, whereas extension of a knee is driven by extensor muscles. **Reciprocal inhibition** is a neural mechanism for simultaneous activation of working muscles with inhibition of antagonist muscles. For example, bending of a knee is driven by the activation of flexor muscles with simultaneous inhibition of extensor muscles. In the spinal cord, stimulatory output from the central nervous system synapses on motor neurons that activate the muscles that power the movement. Simultaneously, the central stimulatory output forms a synapse with an inhibitory interneuron, which inhibits the contraction of antagonist muscles.

CARDIAC MUSCLE

Cardiac Muscle cells are muscle cells in the heart that power the ejection of blood into the circulation. Cardiac muscle cells are striated muscle cells that are interconnected via **intercalated discs** (Fig. 4.18). Desmosomes in the intercalated disc provide strong mechanical connections between cardiac muscle cells. Gap junctions, nonspecific ion channels in the intercalated disc, provide electrical synchronization of cardiac muscle cells.

Being striated muscle cells, cardiac muscle cells are regulated by the Ca^{2+}-troponin-tropomyosin system for contraction. Unlike skeletal muscle cells, cardiac muscle cells rely on Ca^{2+} influx through Ca^{2+} channels on the cell membrane in addition to intracellular Ca^{2+} release from sarcoplasmic reticulum for increasing intracellular $[Ca^{2+}]$ (Fig. 4.19). During cardiac muscle contraction, Ca^{2+} influx through L-type Ca^{2+} channels on the cell membrane causes a moderate increase in intracellular Ca^{2+}, which functions as a trigger for stimulating the opening of ryanodine Ca^{2+} channels on the sarcoplasmic reticulum for Ca^{2+} release by a process known as Ca^{2+}-induced Ca^{2+} release. During cardiac muscle relaxation, some Ca^{2+} is extruded from cytoplasm to the extracellular fluid by Na^+/Ca^{2+} exchange on the cell membrane, and some Ca^{2+} is pumped back to the sarcoplasmic reticulum by Ca^{2+}-ATPase on the sarcoplasmic reticular membrane (Fig. 4.19). The dependence of cardiac muscle contraction on Ca^{2+} influx from extracellular fluid implies that extracellular $[Ca^{2+}]$ is critical for the contractile function of the heart.

SMOOTH MUSCLE

Smooth Muscle cells regulate the dimension of hollow organs—for example, airways, blood vessels, gastrointestinal tract, and uterus. Airway smooth muscle cells regulate the dimension of airways, thereby regulating airway resistance and lung ventilation. Similarly, vascular smooth muscle cells regulate the dimension of blood vessels, thereby regulating vascular resistance and organ blood flow (Fig. 4.20). Gastrointestinal smooth muscle cells regulate the contractility of the gastrointestinal tract, thereby regulating the movement of food down the gastrointestinal tract. Smooth muscle cells appear smooth without striations under light microscopy, because actin and myosin filaments are present at relatively low density in smooth muscle cells. Actin filaments are anchored to dense plaques on the cell membrane and dense bodies in the cytoplasm (Fig. 4.21). The mesh work of actin and myosin filaments enable smooth muscle cells to contract at multiple directions.

Unlike striated muscle cells, smooth muscle cells do not contain troponin—the Ca^{2+}-binding protein that regulates thin filament activation. Ca^{2+} regulates smooth muscle contraction by forming the active enzyme complex, Ca^{2+}_{4}-calmodulin-**myosin light chain kinase** (MLCK), for catalyzing myosin light chain phosphorylation—a necessary step for the activation of myosin crossbridges (Fig. 4.22). The reverse reaction—myosin light chain dephosphorylation—is catalyzed by **myosin light chain phosphatase** (MLCP). The MLCK/MLCP activity ratio determines the level of myosin light chain phosphorylation and smooth muscle activation. During smooth muscle contraction, an increase in MLCK/MLCP ratio is caused by Ca^{2+}/calmodulin-dependent MLCK activation and protein kinase-dependent MLCP inhibition, as examined in the next paragraph.

Different smooth muscle types utilize different membrane signaling mechanisms for activation. For example, gastrointestinal smooth muscle cells utilize membrane depolarization and action potentials for activation. In comparison, airway and vascular smooth muscle cells utilize G protein-coupled receptor signaling mechanisms for activation. For example, as shown in Fig. 4.23, G protein-coupled receptor-mediated activation of the enzyme **phospholipase C** (PLC) results in the breakdown of the membrane phospholipid, phosphatidylinositol bisphosphate (PIP_2), into two second messengers—inositol trisphosphate (IP_3) and diacylglycerol (**DAG**). IP_3 activates IP_3-receptor calcium channels on the sarcoplasmic reticular membrane, resulting in the release of Ca^{2+} from the sarcoplasmic reticulum to the cytoplasm. DAG activates protein kinase C (**PKC**), which serves two functions—activation of Ca^{2+} channels in the cell membrane to increase Ca^{2+} influx and inhibition of MLCP. Ca^{2+}/calmodulin-dependent MLCK activation and PKC-dependent MLCP inhibition together contribute to the increase in MLCK/MLCP ratio for increasing myosin light chain phosphorylation and activating smooth muscle contraction.

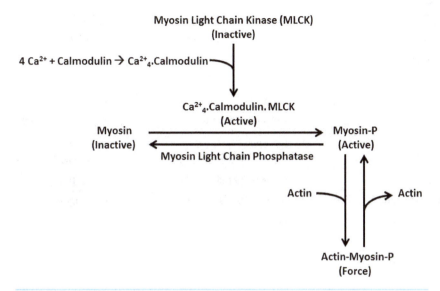

Fig. 4.22—Ca²⁺-dependent Regulatory Mechanism of Smooth Muscle Contraction. Smooth muscle contractions are typically activated by G-protein-coupled receptors and regulated by Ca²⁺, calmodulin-dependent myosin light chain kinase (MLCK).

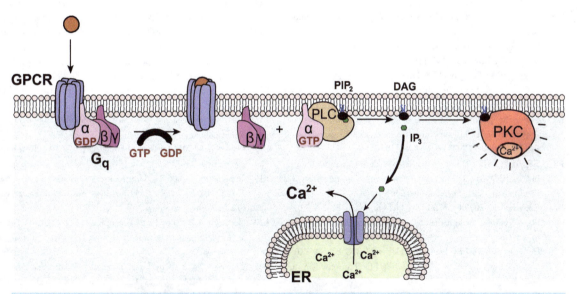

Fig. 4.23—G-protein-coupled Receptor (GPCR) Linked to Phospholipase C (PLC) Activation is a Common Mechanism for Smooth Muscle Activation. A ligand (e.g. neurotransmitter) binds to the GPCR, resulting in the activation of PLC, which catalyzes the hydrolysis of PIP_2 to form two second messengers – inositol trisphosphate (IP_3) and diacylglycerol (DAG). IP_3 activates calcium channels on the endoplasmic reticulum (ER) to induce the release of Ca²⁺ into the cytoplasm for activating smooth muscle contraction. DAG activates protein kinase C (PKC) to activate a cascade of protein phosphorylation for smooth muscle contraction.

KEY TERMS

- alpha-gamma coactivation
- alpha motor neuron
- botox
- cardiac muscle
- concentric contraction
- crossbridge cycle
- DAG
- eccentric contraction
- excitation-contraction coupling
- gamma motor neuron

- Golgi tendon reflex
- intercalated disc
- IP3
- isometric contraction
- isotonic shortening
- muscle fiber types
- myosin light chain kinase
- myosin light chain phosphatase
- neuromuscular junction
- phospholipase C

- PKC
- reciprocal inhibition
- sarcomere
- skeletal muscle
- smooth muscle
- stretch reflex
- striated muscle
- tropomyosin
- troponin

IMAGE CREDITS

5

ENDOCRINE PHYSIOLOGY

ENDOCRINE SYSTEM

Hormones are powerful molecules. Thyroid hormones are essential for brain development in newborns. Cortisol is essential for survival by maintaining normal blood glucose and blood pressure. Hormones are secreted by ductless endocrine glands into the surrounding interstitium for diffusion into the circulation. Protein hormones—for example, growth hormone and insulin—are water-soluble, exist as free molecules in circulation, and typically bind to receptors on the cell membrane of target cells. Many hormones are not proteins. Thyroid hormones are amines. Cortisol, testosterone, and estrogen are steroids. Amine and steroid hormones are lipid-soluble, bound to carrier proteins in circulation, and typically bind to cytoplasmic or nuclear receptors in target cells, but some steroid hormones also bind to receptors on the cell membrane of target cells.

Fig. 5.1A shows the major endocrine glands in the human body. This chapter focuses mostly on the pituitary gland, thyroid gland, adrenal gland, pancreas, and reproductive endocrine glands. The pineal gland and the thymus are regulators of circadian rhythm and immune function, respectively. Functions of the pineal gland and thymus are not covered in this chapter.

LEARNING OBJECTIVES

1. **Hypothalamic–Pituitary–Adrenal (HPA) Axis.** Identify the hormones along the HPA axis, and compare and contrast adrenal deficiency (Addison's disease) and hypercortisolism (Cushing syndrome) in terms of causes and pathophysiological features.

2. **Hypothalamic–Pituitary–Thyroid (HPT) Axis.** Identify the hormones along the HPT axis; describe the mechanism of thyroid hormone synthesis by thyroid follicular cells; compare and contrast hyperthyroidism (Graves' disease) and hypothyroidism (Hashimoto's disease) in terms of causes and pathophysiological features.

3. **Hypothalamic–Pituitary–Gonad (HPG) Axis.** Compare and contrast the HPG axis in males and females in terms of hormones along the axis and their reproductive functions.

4. **Hypothalamic–Pituitary–Liver (HPL) Axis.** Identify the hormones along the HPL axis, and compare and contrast the pathophysiological features of excessive growth hormone secretion in adolescents and adults.

5. **Hypothalamic Regulation of Prolactin Secretion.** Identify the hypothalamic neurotransmitter and hypothalamic hormone that regulate prolactin secretion by the anterior pituitary, and discuss the potential side effect of some antidepressants on prolactin secretion.

6. **Parathyroid Hormone and Vitamin D.** Describe the direct effect and indirect effect (via vitamin D activation) of parathyroid hormone on the regulation of plasma $[Ca^{2+}]$ by bone, gastrointestinal tract, and kidney.

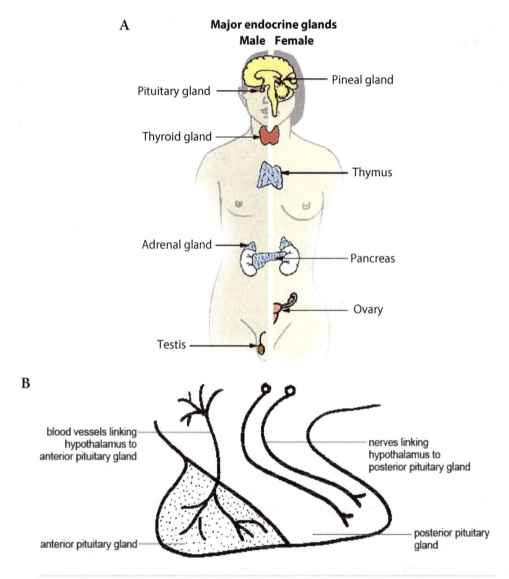

Fig. 5.1—A. Major endocrine glands in the human body.   **B.** Anterior and posterior pituitary glands.

The **Pituitary Gland** consists of anterior and posterior pituitary glands (Fig. 5.1B). The posterior pituitary gland contains projections from endocrine cells residing in the hypothalamus. Posterior pituitary hormones are synthesized in the hypothalamus and released in the posterior pituitary gland. In comparison, the anterior pituitary gland contains endocrine cells that synthesize hormones in the anterior pituitary gland. The anterior pituitary gland is part of the hypothalamic-pituitary-endocrine gland system, consisting of a cascade of hypothalamic hormone(s) (HH), anterior pituitary hormone(s) (PH), and target hormone(s) (TH) released by a peripheral endocrine gland, as shown in the following scheme:

$$\text{Hypothalamus} \rightarrow \text{HH} \rightarrow \text{Anterior Pituitary} \rightarrow \text{PH} \rightarrow \text{Endocrine Gland} \rightarrow \text{TH}$$
$$\text{(Portal Circulation)}$$

A portal circulation carries hypothalamic hormone from the capillary network in the hypothalamus to the capillary network in the anterior pituitary, where the hypothalamic hormone (HH) stimulates or inhibits the release of pituitary hormone (PH) from the anterior pituitary to the general circulation. The pituitary hormone then stimulates the release of a target hormone (TH) from a peripheral endocrine gland. There are five hypothalamic-pituitary-endocrine gland systems.

THE ADRENAL GLAND

Hypothalamic-Pituitary-Adrenal Axis. Fig. 5.2 shows the cascade of three hormones in this axis. The hypothalamus secretes a corticotrophin-releasing hormone (CRH), which stimulates the release of adrenocorticotrophic hormone (ACTH) by the anterior pituitary gland to the general circulation. ACTH then stimulates the release of **cortisol**, a **glucocorticoid**, from the adrenal cortex. Cortisol is named glucocorticoid, because cortisol is essential for maintaining normal plasma [glucose]. Cortisol is essential for survival, because cortisol is required for maintaining normal blood pressure and plasma glucose concentration. Cortisol is a suppressor of the immune response. Synthetic glucocorticoids are used clinically to treat inflammatory diseases.

As shown in Fig. 5.2, by negative feedback, cortisol suppresses ACTH production by the anterior pituitary gland and CRH production by the hypothalamus. Understanding this negative feedback is important for diagnosis of adrenal cortical diseases, as illustrated in Fig. 5.3. In primary adrenal insufficiency caused by the inability of adrenal cortex to produce cortisol, the hypothalamus will increase CRH secretion and the anterior pituitary will increase ACTH secretion in response to low plasma [cortisol], as shown in Fig. 5.3 **(top row)**. In secondary adrenal deficiency caused by the inability of the anterior pituitary gland to produce ACTH, the adrenal cortex will decrease cortisol secretion, but the

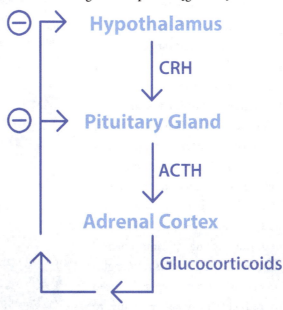

Fig. 5.2—Hypothalamic-pituitary-adrenal Axis. The hypothalamic hormone, corticotrophic releasing hormone (CRH), stimulates the release of adrenocorticotrophic hormone (ACTH) by the anterior pituitary gland. ACTH stimulates the release of cortisol, a glucocorticoid, by the adrenal cortex. By negative feedback, cortisol inhibits the releases of CRH and ACTH.

hypothalamus will increase CRH secretion in response to low plasma [cortisol], as shown in Fig. 5.3 (**middle row**). In tertiary adrenal deficiency caused by the inability of hypothalamus to secrete CRH, the anterior pituitary will decrease ACTH secretion and the adrenal cortex will decrease cortisol secretion, as shown in Fig. 5.3 (**bottom row**).

Addison's disease, a type of primary adrenal insufficiency, is characterized by abnormally low plasma glucose concentration, low blood pressure, and fatigue. Taking glucocorticoid drugs can cause inhibition of ACTH secretion by the anterior pituitary gland and inhibition of cortisol secretion by the adrenal cortex. Abrupt cessation in taking glucocorticoid drugs can lead to secondary adrenal insufficiency because the inhibited adrenal gland is unable to maintain normal plasma cortisol concentration. Patients taking glucocorticoid drugs typically need to reduce the dose gradually to allow the adrenal glands to restore a normal level of cortisol synthesis and secretion.

Cushing's syndrome, caused by ACTH-secreting tumors in

Organ of insufficiency	CRH	Hormones ACTH	Cortisol
Adrenal gland (Primary adrenal insufficiency)	High	High	Low
Anterior pituitary (Secondary adrenal insufficiency)	High	Low	Low
Hypothalamus (Tertiary adrenal insufficiency)	Low	Low	Low

Fig. 5.3—Primary and Secondary Adrenal Insufficiency. A. Primary adrenal insufficiency is caused by the inability of adrenal glands to produce cortisol. In response to low plasma [cortisol], the hypothalamus and anterior pituitary increase the release of CRH and ACTH, respectively. **B.** Secondary adrenal deficiency can be caused by the inability of hypothalamus to produce CRH, resulting in low levels of ACTH and cortisol. Tertiary adrenal deficiency can be caused by the inability of the anterior pituitary gland to produce ACTH, resulting in low cortisol, which causes an increase in CRH by negative feedback.

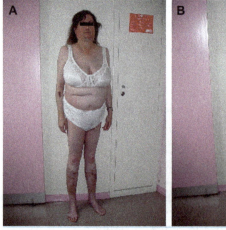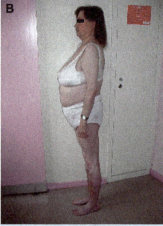

Fig. 5.4. Cushing's syndrome—caused by excessive production of cortisol—is characterized by centripetal fat depositing with truncal obesity-contrasting with the muscular atrophy of the thighs and legs. (From: Bertagna et al. Best Practice Res Clin Endocrinol Metabolism 23: 610, 2009)

the anterior pituitary gland or cortisol–secreting tumors in the adrenal gland, is characterized by abnormally high plasma concentrations of cortisol and glucose, abnormal distribution of fat to the face and abdomen, and loss of muscle in thighs and legs, as shown in Fig. 5.4.

In addition to cortisol, the adrenal cortex also secretes aldosterone and dehydroepiandrosterone (DHEA). **Aldosterone** is named **mineralocorticoid**, because aldosterone regulates Na$^+$ reabsorption in the kidneys. **DHEA** is an **androgen**, because DHEA is a precursor for testosterone synthesis.

The functions of aldosterone and testosterone will be examined in the chapters on renal and reproductive physiology.

THE THYROID GLAND

Hypothalamic-Pituitary-Thyroid Axis. Fig. 5.5 shows the cascade of hormones in this axis. The hypothalamus secretes thyrotropin-releasing hormone (TRH), which stimulates the anterior pituitary to secrete thyroid-stimulating hormone (TSH) into the general circulation. TSH then stimulates the thyroid gland to secrete two thyroid hormones—triiodothyronine (T3) and thyroxine (T4). By negative feedback, T3 and T4 inhibit the TRH secretion by the hypothalamus and TSH secretion by the anterior pituitary. Thyroid

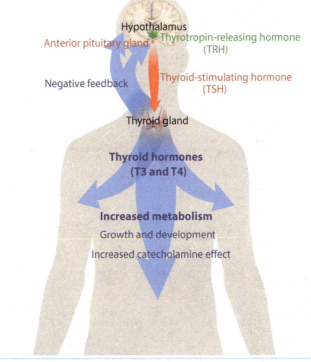

Thyroid system

Hypothalamus
Anterior pituitary gland
Thyrotropin-releasing hormone (TRH)
Negative feedback
Thyroid-stimulating hormone (TSH)
Thyroid gland
Thyroid hormones (T3 and T4)
Increased metabolism
Growth and development
Increased catecholamine effect

Fig. 5.5. Hypothalmus-pituitary-thyroid axis.

hormones are essential for the regulation of brain development in children and basal metabolic rate in adults.

Fig. 5.6A shows the mechanism of thyroid hormone synthesis in a thyroid gland, consisting of a central reservoir of follicular colloid and surrounding follicular cells. Follicular cells sequester iodine (I⁻) from the blood via the Na⁺/I⁻ co-transporter and release I⁻ into the follicular colloid via the anion transporter pendrin. Follicular cells also synthesize and release thyroglobulin into the follicular colloid. The enzyme thyroid peroxidase present in the follicular colloid catalyzes the oxidation of I⁻ to reactive iodine (I°), iodination of tyrosine residues of thyroglobulin, and conjugation of iodotyrosyl residues to form T3 and T4 as part of the iodinated thyroglobulin molecule. The last step of thyroid hormone synthesis involves endocytosis of iodinated thyroglobulin by follicular cells, release of T3 and T4 from thyroglobulin by proteolysis, and diffusion of T3 and T4 into the circulation.

Graves' disease is an autoimmune disease of hyperthyroidism. In Graves' disease, autoimmune antibodies stimulate TSH receptors in the thyroid gland, causing excessive secretion of thyroid hormones into the circulation and enlargement of the gland as goiter in the neck (Fig. 5.6B). Patients having hyperthyroidism have abnormally high plasma concentrations of thyroid hormones, abnormally low plasma concentration of TSH, abnormally high heart rate, abnormally high basal metabolic rate, and weight loss.

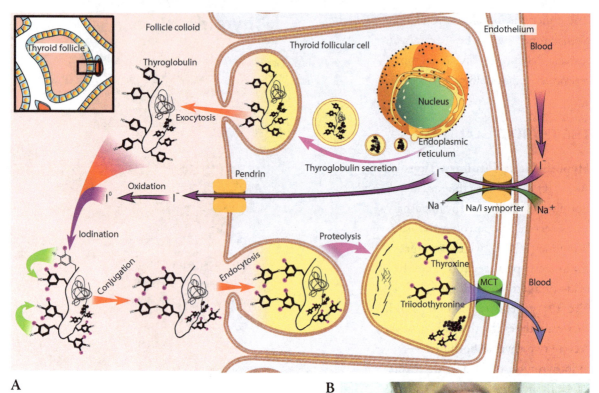

A

Hashimoto's disease is an autoimmune disease of hypothyroidism. In Hashimoto's disease, immune cells and autoimmune antibodies destroy thyroid peroxidase and/or thyroglobulin, leading to destruction of the thyroid follicles. Patients having hypothyroidism have abnormally low plasma concentrations of thyroid hormones, abnormally high plasma concentration of TSH, and abnormally low metabolic rate.

B

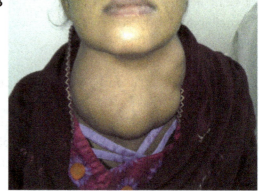

Fig. 5.6—A. Thyroid hormone synthesis. **B.** Goiter formation in Graves' disease.

REPRODUCTIVE HORMONES

Hypothalamic-Pituitary-Gonad Axis. Fig. 5.7 shows the cascade of hormones in this axis. As shown in Fig. 5.7, in both females and males, the hypothalamus secretes gonadotropin-releasing hormone (GnRH), which stimulates the anterior pituitary gland to secrete two hormones—follicle stimulating hormone (FSH) and luteinizing hormone (LH). FSH stimulates germ cell production, whereas LH stimulates reproductive hormones secretion.

The female reproductive endocrine system is shown in Fig. 5.7A. FSH stimulates egg development (oogenesis), and LH stimulates the secretion of **estrogen** and **progesterone** in the ovaries. By negative feedback, estrogen and progesterone inhibit GnRH secretion by the hypothalamus and

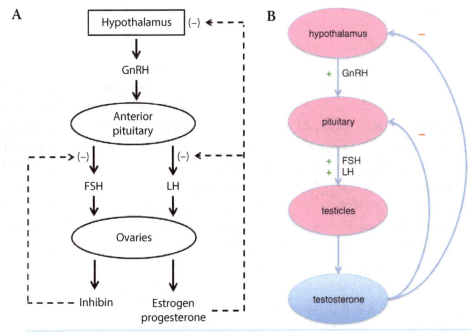

Fig. 5.7—A. Hypothalamus-pituitary-ovarian axis in females and **B.** hypothalamus-pituitary-testis axis in males.

LH secretion by the anterior pituitary. The male reproductive endocrine system is shown in Fig. 5.7B. FSH stimulates sperm development (spermatogenesis), and LH stimulates the secretion of **testosterone** by the testes. Reproductive hormones stimulate secondary sexual characteristics by regulating gene expression in target cells. Reproductive function of the hypothalamic-pituitary-testis axis will be examined in detail in the chapter on reproductive physiology.

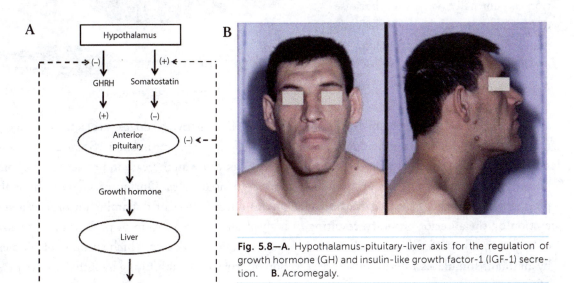

Fig. 5.8—A. Hypothalamus-pituitary-liver axis for the regulation of growth hormone (GH) and insulin-like growth factor-1 (IGF-1) secretion. **B.** Acromegaly.

GROWTH HORMONE

Hypothalamic-Pituitary-Liver Axis. Fig. 5.8A shows the cascade of hormones in this axis. The hypothalamus secretes two hormones—growth hormone releasing hormone (GHRH) and somatostatin. GHRH stimulates, whereas somatostatin inhibits **growth hormone** secretion by the anterior pituitary. Growth hormone then stimulates the liver to secrete insulin-like growth factor-1 (IGF-1), which is the major mediator of bone growth. By negative feedback, IGF-1 inhibits GHRH secretion and stimulates somatostatin secretion by the hypothalamus, and inhibits growth hormone secretion by the anterior pituitary. In adolescents having the growth plate, the growth hormone-IGF-1 axis stimulates bone lengthening. Excessive secretion of growth hormone and/or IGF-1 in well–fed adolescents can lead to gigantism. Children having malnutrition—caused by starvation or inflammatory bowel disease—have retarded growth, because the liver is unable to secrete a normal level of IGF-1, whereas plasma concentration of growth hormone may be abnormally high in these children. Restoration of nutrition usually leads to catch-up of growth in these children. After closing of the growth plate in adults, excessive secretion of growth hormone and/or IGF-1 can lead to acromegaly—thickening of fingers and broadening of the face (Fig. 5.8B).

PROLACTIN

Hypothalamic Regulation of Prolactin Secretion. Fig. 5.9 shows the regulation of prolactin secretion by the anterior pituitary. In nonpregnant females, plasma prolactin concentration is normally very low, because dopamine released by hypothalamic neurons strongly inhibits prolactin secretion by the anterior pituitary. The hypothalamic hormone thyrotropin-releasing hormone (TRH) weakly

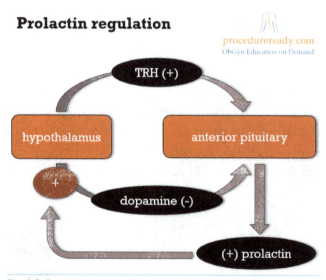

Prolactin regulation

Fig. 5.9. Regulation of prolactin secretion.

stimulates prolactin secretion by the anterior pituitary. By negative feedback, prolactin stimulates the secretion of dopamine and inhibits secretion of TRH by the hypothalamus. Some antidepressants having dopaminergic receptor-blocking activity can cause an increase in prolactin secretion, breast enlargement, and milk production in nonpregnant females. During pregnancy, the high plasma concentration of estrogen overcomes the inhibitory effect of dopamine on prolactin secretion by the anterior pituitary, resulting in high plasma concentration of prolactin and breast enlargement. Breastfeeding stimulates milk production via a reflex, in which suckling of nipples by an infant stimulates anterior pituitary secretion of prolactin by inhibiting hypothalamic release of dopamine.

POSTERIOR PITUITARY HORMONES

Posterior pituitary hormones—oxytocin and antidiuretic hormone—are synthesized by hypothalamic neurons and then released via nerve endings in the posterior pituitary gland. Oxytocin is essential for stimulating uterine smooth muscle contraction during labor and stimulating milk ejection during breastfeeding. The function of oxytocin will be covered in detail in the chapter on reproductive physiology. Antidiuretic hormone is essential for regulating extracellular fluid osmolarity by regulating urinary excretion of water. The function of antidiuretic hormone will also be examined in detail in the chapter on renal physiology. Oxytocin and antidiuretic hormone are considered neurohormones, because they regulate both neurological and physiological functions.

PARATHYROID HORMONE AND VITAMIN D

As shown in Fig. 5.10A, four **parathyroid glands** are embedded within the thyroid gland. Parathyroid hormone is essential for maintaining plasma calcium concentration within physiological range, which is essential for normal function of all organ systems. As shown in Fig. 5.10B, parathyroid hormone increases plasma calcium concentration by direct and indirect mechanisms. Parathyroid hormone directly stimulates calcium release from bone resorption and calcium reabsorption in the kidney. Parathyroid hormone indirectly increases plasma calcium concentration by enhancing the activation of **vitamin D** by hydroxylation in the kidney. Inactive vitamin

A   Thyroid and parathyroid glands

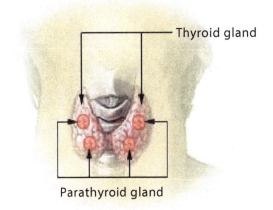

Thyroid gland

Parathyroid gland

B   Calcium regulation

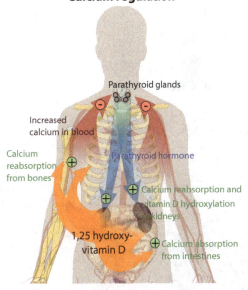

Parathyroid glands

Increased calcium in blood

Calcium reabsorption from bones

Parathyroid hormone

Calcium reabsorption and vitamin D hydroxylation kidneys

1,25 hydroxy-vitamin D

Calcium absorption from intestines

Fig. 5.10—A. Parathyroid gland. **B.** Regulation of plasma [Ca²⁺] by parathyroid hormone.

D is available exogenously from food or endogenously from synthesis by skin cells in response to ultraviolet irradiation. Activation of vitamin D requires hydroxylation at the 25 position in the liver and at the 1 position in the kidney. Active 1,25-dihydroxy-vitamin D is required for intestinal absorption of calcium. Severe vitamin D deficiency in children can lead to bone demineralization and deformation, a disease condition known as rickets. Hyperparathyroidism can lead to excessive bone resorption and osteoporosis.

Gastrointestinal and Pancreatic Hormones. The gastrointestinal tract and pancreas contain endocrine cells. Gastrointestinal hormones regulate gastrointestinal motility, secretion, and appetite. Pancreatic hormones regulate energy metabolism and plasma [glucose]. In-depth discussion of gastrointestinal and pancreatic hormones will be presented in the chapter on gastrointestinal physiology.

KEY TERMS

- Addison's disease
- adrenal gland
- aldosterone
- androgen
- antidiuretic hormone
- cortisol
- Cushing's syndrome
- DHEA
- estrogen
- glucocorticoid
- Graves' disease
- Growth hormone
- Hashimoto's disease
- hypothalamic-pituitary-adrenal axis
- hypothalamic-pituitary-gonad axis
- hypothalamic-pituitary-liver axis
- hypothalamic-pituitary-thyroid axis
- hypothalamus
- mineralocorticoid
- negative feedback
- oxytocin
- parathyroid gland
- parathyroid hormone
- pituitary glands
- progesterone
- prolactin
- testosterone
- thyroid gland
- vitamin D

IMAGE CREDITS

5.1a NCI / NIH / Copyright in the Public Domain.
5.1b Copyright © 2007 by Ruth Lawson - Otago Polytechnic / (CC BY 3.0) at http://commons.wikimedia.org/wiki/File:Anterior_and_posterior_pituitary.jpg.
5.2 Copyright © 2008 by User:Drosenbach / Wikimedia Commons / (CC BY 3.0) at http://commons.wikimedia.org/wiki/File:ACTH_Negative_Feedback.svg.
5.4 Bertagna, Laurence Guignat, Lionel Groussin, and Jérôme Bertherat, from Best Practice & Research Clinical Endocrinology & Metabolism, vol. 23, no. 5, p. 610. Copyright © 2009 by Elsevier Science and Technology. Reprinted with permission.
5.5 Mikael Häggström / Copyright in the Public Domain.
5.6a Mikael Häggström / Copyright in the Public Domain.
5.6b User:Almazi / Wikimedia Commons / Copyright in the Public Domain.
5.7b Copyright © 2010 by User:Boghog / Wikimedia Commons / (CC BY-SA 3.0) at http://commons.wikimedia.org/wiki/File:Hypothalamus_pituitary_testicles_axis.png.
5.8b Copyright © 2008 by Philippe Chanson and Sylvie Salenave / (CC BY 2.0) at http://commons.wikimedia.org/wiki/File:Acromegaly_facial_features.JPEG.
5.9 Copyright © 2014 by User:Procedureready / Wikimedia Commons / (CC BY-SA 3.0) at http://commons.wikimedia.org/wiki/File:Prolactin_regulation.png.
5.10a NCI / NIH / Copyright in the Public Domain.
5.10b Mikael Häggström / Copyright in the Public Domain.

6

CARDIOVASCULAR PHYSIOLOGY

Heart disease is the leading cause of death in developed countries. The cardiovascular system, in concert with the respiratory system, is essential to life on a minute-to-minute basis. This chapter discusses the functions of the heart and vascular system in circulating blood to the lungs for the loading of oxygen and unloading of carbon dioxide, and to the organ systems for the unloading of oxygen and loading of carbon dioxide. In addition to carrying oxygen and carbon dioxide, the circulation is also essential for carrying hormones for regulation, nutrients for consumption, and waste products for disposal.

OVERVIEW OF THE CARDIOVASCULAR SYSTEM

Fig. 6.1 shows the structure of the cardiovascular system. The pump of the cardiovascular system is the heart, which has two chambers—the left heart and the right heart—separated by a septum. In physiology, the cardiovascular system is typically shown as positioned inside a person facing the reader—i.e., the left heart is shown

LEARNING OBJECTIVES

1. **Cardiac Electrophysiology and Electrocardiogram (ECG).** Describe the electrophysiological mechanisms of heartbeat generation, including the ionic basis of pacemaker potential and cardiac action potential; illustrate the electrocardiographic connections for recording a 12-lead ECG; interpret a normal 12-lead ECG.

2. **Cardiac Cycle and Determinants of Cardiac Output.** Define the four phases of cardiac cycle, and explain how electrical events of the heart (as marked by ECG), together with cardiac valves, regulate left ventricular volume, left ventricular pressure, aortic pressure, and atrial pressure during the cardiac cycle; discuss the mechanisms by which four determinants regulate cardiac output; describe typical measurements of the four determinants of cardiac output.

3. **Vascular System and Arterial Pressure.** Compare and contrast the structure and function of the four major segments of the vascular system; and describe the measurement of diastolic and systolic arterial blood pressure using a manometer.

4. **Control of Organ Blood Flow.** Compare and contrast neural and local control of organ blood flow in terms of mechanisms and organ specificity.

5. **Endothelium-Derived Relaxing Factor (EDRF), Renin-Angiotensin-Aldosterone System, and Baroreceptor Reflex.** Discuss the mechanism of EDRF (nitric oxide) production by endothelial cells and mechanism of nitric oxide-induced vascular smooth muscle relaxation; and compare and contrast renin–angiotensin–aldosterone system and baroreceptor reflex in terms of signaling pathway and physiological function.

6. **Capillary Fluid Filtration and Reabsorption.** Compare and contrast the roles of hydrostatic pressure and protein osmotic pressure in the regulation of fluid movement across a capillary wall.

7. **Integrative Cardiovascular Physiology and Pathophysiology.** Compare and contrast the effects of hemorrhage and systolic heart failure on the cardiovascular system in terms of mechanisms and compensatory responses of the cardiovascular system, and explain the mechanism of exercise-induced increase in cardiac output.

on the right side, and the right heart is shown on the left side of the diagram. As shown in Fig. 6.1, the left heart is colored red to indicate that the left atrium receives oxygenated blood from the lungs via the pulmonary veins and the left ventricle pumps oxygenated blood to organs via the aorta. At each organ other than the lungs, after the unloading of oxygen from blood to the organs, deoxygenated venous blood leaves the organ via a systemic vein to return to the right **atrium** via the inferior and superior vena cava. This section of the cardiovascular system, which begins in the left **ventricle** and ends in the right atrium, is termed "**systemic circulation.**" The right heart is shown in blue to indicate that the right atrium receives deoxygenated blood from organ systems via the inferior and superior vena cava. The right ventricle pumps the deoxygenated blood to the lungs via the pulmonary arteries. After the loading of oxygen from air in the lungs to the blood, oxygenated pulmonary venous blood returns to the left atrium via the pulmonary veins. This section of the cardiovascular system, which begins in the right ventricle and ends in the left atrium, is termed "**pulmonary circulation.**"

At this point, it is important to address the nomenclature of blood vessels as arteries and veins. Arteries are defined anatomically as blood vessels leaving the heart; for example, the aorta is an artery, because the aorta carries blood away from the left ventricle. Similarly, pulmonary arteries carry blood away from the right ventricle. Veins are defined anatomically as blood vessels returning to the heart; for example, the inferior vena cava is a vein, because the inferior vena cava carries blood into the

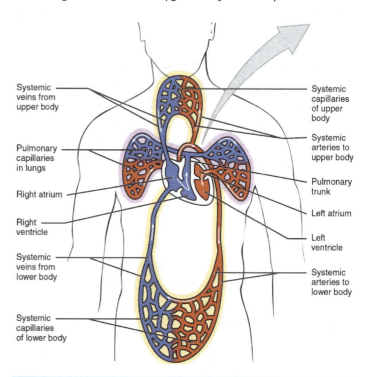

Fig. 6.1. Overview of the circulatory system.

right atrium. Similarly, pulmonary veins carry blood into the left atrium. Arteries and veins may carry oxygenated or deoxygenated blood, depending on whether the blood vessels are located in the systemic or pulmonary circulation. As shown in Fig. 6.1, systemic arteries carry oxygenated blood (red) and systemic veins carry deoxygenated blood (blue). In comparison, pulmonary arteries carry deoxygenated blood (blue) and pulmonary veins carry oxygenated blood (red).

Systemic and pulmonary circulations are connected in series, as shown in the following scheme:

LV → Systemic Circulation → RA → RV → Pulmonary Circulation → LA

↑_____↓

As shown in the above scheme, systemic circulation begins in the left ventricle (LV) and ends in the right atrium (RA). The right atrium (RA) empties blood into the right ventricle (RV). The pulmonary circulation begins in the right ventricle (RV) and ends in the left atrium (LA). The left atrium (LA) empties into the left ventricle (LV). One implication of this architecture of the cardio-vascular system is that blood volume is distributed between the systemic circulation and pulmonary circulation. To establish steady-state blood volumes in the systemic and pulmonary circulations, the left and right cardiac outputs must be equal. In principle, any inequality between left and right cardiac outputs will result in the accumulation of blood volume exclusively in one circulation. This does not occur in normal physiology, because an intrinsic mechanism of the heart—Frank-Starling mechanism—constantly adjusts cardiac output of each ventricle in proportion to venous return to each ventricle, thereby equalizing right and left cardiac output. The Frank-Starling mechanism will be covered in detail later in this chapter.

For both systemic circulation and pulmonary circulation, arterial blood pressure is much higher than venous blood pressure. Arterial **blood pressure** is higher than venous pressure, because the ventricle pumps blood directly into the arterial system, whereas blood pressure is dissipated dur-ing blood flow through resistance vessels in organs into the venous system. In both systemic and pulmonary circulation, blood flow is driven by the arterial-venous pressure difference. As cited previ-ously, left cardiac output to the systemic circulation equals right cardiac output to the pulmonary circulation. In comparison, systemic arterial pressure is higher than pulmonary arterial pressure because systemic vascular resistance is higher than pulmonary **vascular resistance**. These differences between systemic circulation and pulmonary circulation are summarized as follows:

Left Cardiac Output to Systemic Circ. = Right Cardiac Output to Pulmonary Circ.

Systemic Arterial Pressure > Pulmonary Arterial Pressure

Systemic Vascular Resistance > Pulmonary Vascular Resistance

The systemic circulation is considered a high pressure-high resistance circuit, whereas the pulmo-nary circulation is considered a low pressure-low resistance circuit. The left ventricular wall is much thicker than the right ventricular wall, because the left ventricle pumps blood against a relatively high systemic arterial pressure, whereas the right ventricle pumps blood against a relatively low pulmonary arterial pressure.

Structure of the Heart. As shown in Fig. 6.2, there are four valves and four chambers in the heart. Cardiac valves are essential for controlling unidirectional blood flow from the atria to the

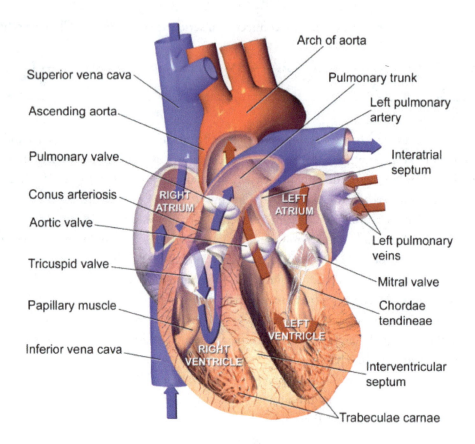

Arch of aorta

Superior vena cava

Ascending aorta

Pulmonary valve

Conus arteriosis

Aortic valve

Tricuspid valve

Papillary muscle

Inferior vena cava

Pulmonary trunk

Left pulmonary artery

Interatrial septum

RIGHT ATRIUM

LEFT ATRIUM

Left pulmonary veins

Mitral valve

Chordae tendineae

LEFT VENTRICLE

RIGHT VENTRICLE

Interventricular septum

Trabeculae carnae

Sectional Anatomy of the Heart

Fig. 6.2. Structure of the heart.

ventricles and from the ventricles to the arteries. For example, the **aortic valve** controls unidirectional blood flow from the left ventricle to the aorta by opening only when left ventricular pressure is higher than aortic pressure. The aortic valve closes when left ventricular pressure is lower than aortic pressure during ventricular filling to prevent regurgitation of blood from the aorta to the left ventricle. Similarly, the **pulmonary valve** controls unidirectional blood flow from the right ventricle to the pulmonary arteries. The left atrioventricular valve—also known as the **mitral valve**—controls unidirectional blood flow from the left atrium to the left ventricle by opening only during ventricular filling. The mitral valve closes during ventricular ejection into the circulation to prevent regurgitation of blood from the left ventricle to the left atrium. Similarly, the right atrioventricular valve—also known as the **tricuspid valve**—controls unidirectional blood flow from the right atrium to the right ventricle. In addition, venous and lymphatic valves control unidirectional blood flow and lymph flow through the circulation.

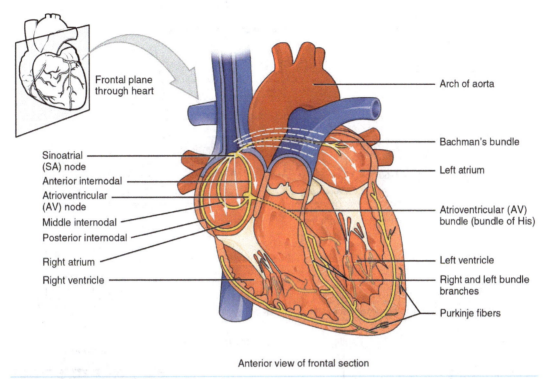

Frontal plane
through heart

Arch of aorta

Bachman's bundle

Sinoatrial
(SA) node

Left atrium

Anterior internodal

Atrioventricular
(AV) node

Atrioventricular (AV)
bundle (bundle of His)

Middle internodal

Posterior internodal

Right atrium

Left ventricle

Right ventricle

Right and left bundle
branches

Purkinje fibers

Anterior view of frontal section

Fig. 6.3. Sinoatral node (pacemaker) and conductive system in the heart.

CARDIAC ELECTROPHYSIOLOGY AND ELECTROCARDIOGRAM

Heartbeat Generation and Conduction. The heart contains an intrinsic pacemaker system that enables the heart to beat without any input from the nervous system. The presence of an intrinsic pacemaker system in the heart is evident in cardiac transplantation, when a transplanted heart is capable of beating at an intrinsic rate in complete separation from the autonomic nervous system. It is important to note that the autonomic nervous system, although unnecessary for heart rate generation, is capable of modulating heart rate by altering pacemaker activity. Fig. 6.3 shows the location of the pacemaker and conductive system in the heart. The **sinoatrial node**, situated in the right atrium, is the cardiac pacemaker, which spontaneously generates action potentials at regular time intervals. Sinoatrial action potentials are conducted directly to atrial cardiac muscle cells, but cannot be conducted directly from atria to ventricles, because the fibrous cardiac skeleton for anchoring the four cardiac valves physically separates the two atria from the two ventricles. Sinoatrial action potentials are conducted indirectly to ventricular cardiac muscle cells through the **atrioventricular node, bundle of His**, right and left bundle branches, and the **purkinje fibers.** Conduction of action potentials through the atrioventricular node is slow, which causes a time delay between atrial depolarization and ventricular depolarization. The atrioventricular delay is functionally significant in allowing active filling of the ventricles by atrial contraction before ventricular ejection of blood into the circulation. The fast conduction of action potentials in the conductive system downstream

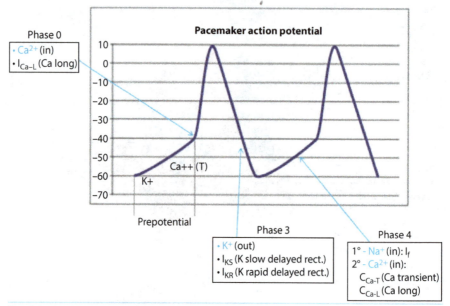

Fig. 6.4—Pacemaker potential of the sinoatrial node.

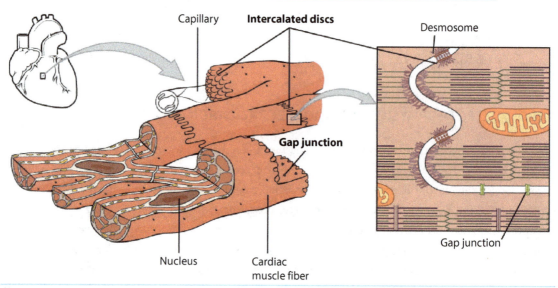

Fig. 6.5. Intercalated disc between cardiac muscle cells.

from atrioventricular node enables synchronized contraction of the two ventricles. It is noteworthy that cells in the conductive system are also capable of generating action potentials spontaneously. The sinoatrial node dominates the control of heart rate, because it generates action potentials at the highest frequency. Cells in the conductive system can function as a safeguard when the sinoatrial node fails to generate action potentials, and function as a filter when the sinoatrial node is generating action potentials at exceedingly high frequencies.

Ionic Basis of Pacemaker Action Potential. Fig. 6.4 shows the spontaneous generation of action potentials by the sinoatrial node. The spontaneous depolarization phase (phase 4)—essential for depolarizing the membrane toward the threshold for action potential generation—is caused mostly by Na^+ influx (I_f) through "funny" channels. These channels are labeled "funny," because they increase their open probability in response to hyperpolarization instead of depolarization. Funny channels are formally known as hyperpolarization-activated, cyclic nucleotide-gated (**HCN**) channels, because they are activated by cyclic AMP. Another contributor to the spontaneous depolarization phase is Ca^{2+} influx through T and L-type calcium channels (I_{Ca-T} and I_{Ca-L}). A sinoatrial nodal cell fires an action potential when membrane depolarization exceeds a critical threshold. The depolarization phase 0 of the action potential is caused by Ca^{2+} influx through L-type voltage-gated calcium channels (I_{Ca-L}), and the repolarization phase 3 is caused by the K^+ efflux through slow and rapid delayed rectifying, voltage-gated potassium channels (I_{KS} and I_{KR}). Phases 1 and 2 are missing in the sinoatrial nodal action potential, because the numbering of phases is derived from the cardiac action potential, where all phases 0 to 4 are present. Phases 1 and 2 correspond to the transient repolarization and plateau depolarization in **cardiac action potential**, which are absent in the sinoatrial action potential.

Sinoatrial action potentials are conducted rapidly to atrial cardiac muscle cells, but slowly to ventricular cardiac muscle cells through the atrioventricular node. Within the atria and ventricles, action potentials are conducted rapidly through gap junctions at intercalated discs between cardiac muscle cells (Fig. 6.5), thereby enabling synchronized contraction of the two atria as one unit and synchronized contraction of the two ventricles as a separate unit. Synchronized contraction of atrial cardiac muscle cells is essential for the filling of ventricles by atrial contraction. Synchronized contraction of ventricular cardiac muscle cells is essential for the ventricular ejection of blood into the circulation. Cardiac arrhythmia—the loss of electrical synchronization of cardiac muscle cells—can lead to decreases in cardiac output and blood pressure, and death. Intercalated discs also contain desmosomes and tight junctions for mechanically strong connections between cardiac muscle cells. Mechanical integrity in all parts of the ventricle is essential for the development of high ventricular pressure necessary for the ejection of blood into the circulation. Mechanical weakness in any part of the ventricular wall can lead to protrusion of the wall (aneurysm) during ventricular contraction, and reduces the efficiency of the ventricle in producing pressure and cardiac output.

Cardiac Action Potential. Cardiac muscle cells in the relaxed state have a stable resting membrane potential. Action potentials in cardiac muscle cells are normally triggered by supra-threshold depolarization that is caused by action potentials from conductive system or neighboring cardiac muscle cells. Fig. 6.6A shows the four phases of a cardiac action potential. Phase 4 represents the stable resting membrane potential in a relaxed cardiac muscle cell, which is dominated by K^+ efflux (I_{K1}) through inward rectifying potassium channels. A supra-threshold depolarization triggers an upstroke depolarization (Phase 0), which is caused by Na^+ influx (I_{Na}) through fast voltage-gated sodium channels. The following brief repolarization (Phase 1) is caused by K^+ efflux ($I_{to}1,2$) through transient outward potassium channels. The long duration (200–250 msec) of plateau depolarization (phase 2) is maintained by Ca^{2+} influx through L-type calcium channels against a relatively small K^+

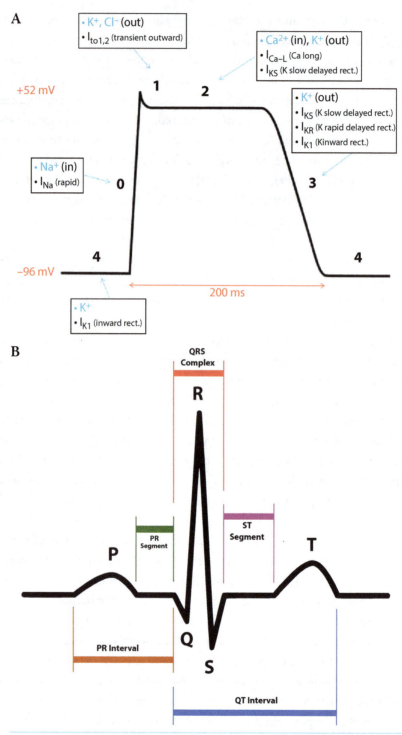

Fig. 6.6—A. Action potential in a ventricular cardiac muscle cell, showing the resting potential (phase 4), upstroke of ventricular depolarization (phase 0), transient repolarization (phase 1), sustained plateau depolarization (phase 2), and repolarization (phase 3).   **B.** Electrocardiogram showing that the QRS is caused by ventricular depolarization and T wave is caused by ventricular repolarization.

efflux through the slow delayed rectifying potassium channels. The long duration (200–250 msec) of plateau depolarization (phase 2) is a unique characteristic of cardiac action potentials. For comparison, the duration of action potentials in motor neurons and skeletal muscle cells is only 1 msec. Phase 3—the repolarization phase—is caused by K^+ efflux through slow delayed rectifying (I_{KS}), rapid delayed rectifying (I_{FR}), and inward rectifying (I_{K1}) potassium channels.

Potassium channels play a critical role in the progression of cardiac action potentials. In patients having long QT syndrome, genetic mutations of potassium channels can lead to delayed opening of potassium channels for repolarization and abnormal lengthening of the cardiac action potential, thereby increasing the risk of developing cardiac arrhythmia.

Basic Principles of the Electrocardiogram. When the atria and ventricles undergo synchronous depolarization and repolarization as two giant cells, they generate electrical currents that can be detected as changes in voltage on the body surface. An electrocardiogram (ECG or EKG) is a recording of electrical activities of the heart by measuring the potential difference (voltage) between two points on the body surface. An electrocardiograph has a positive terminal and a negative terminal. Specific electrocardiographic leads are defined by the attachment sites of the positive and negative ECG terminals on the body surface. For example, Lead II ECG is recorded by attaching the negative terminal to the right arm and the positive terminal to the left leg. Fig. 6.6B shows the P wave, QRS complex, and T wave in a normal Lead II ECG, where the y-axis is voltage and the x-axis is time. The most prominent wave in an electrocardiogram is typically the QRS complex, which is caused by ventricular depolarization (phase 0) of the ventricular cardiac action potential (compare Figs. 6.6A and 6.6B). The wave immediately after the QRS complex, T wave, is caused by ventricular repolarization (Phase 3) of the ventricular cardiac action potential (compare Figs. 6.6A and 6.6B). The wave immediately before the QRS complex, P wave, is caused by atrial depolarization of the atrial cardiac action potential. Atrial repolarization is usually not visible in an electrocardiogram. In a normal electrocardiogram, each P wave is followed by a QRS complex, because each atrial depolarization is followed by a ventricular depolarization. In patients having atrioventricular block, multiple atrial depolarizations (P waves) are generated before each ventricular depolarization (QRS wave). For example, in 2:1 block, two P waves are generated before each QRS complex. In a complete heart block, the occurrence of P waves and QRS complexes become independent of each other. In a normal electrocardiogram, voltage in the S-T segment should be zero because all cardiac muscle cells in the ventricle should be in the depolarization state. The appearance of voltage in the S-T segment—known as S-T elevation—is an indicator of nonuniform depolarization of the ventricle, which is often associated with hypoxia in the heart.

Time intervals between electrocardiographic waves are important indicators of the duration of specific electrical events in the heart. The PR interval between atrial depolarization and ventricular depolarization represents atrioventricular conduction time. The QT time interval between ventricular depolarization and ventricular repolarization represents the duration of the ventricular cardiac action potential and an estimate of ventricular contraction time. The R-R interval between two consecutive ECGs represents the duration of one cardiac cycle.

A

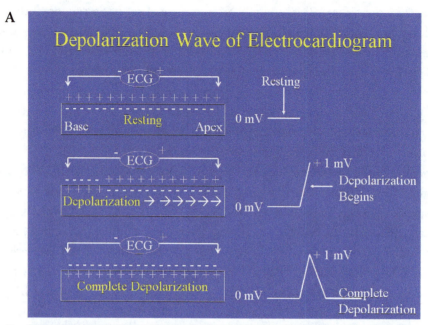

B

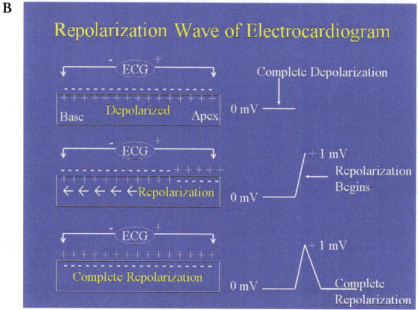

Fig. 6.7. Direction of **A.** depolarization and **B.** repolarization waves of the ventricle. Note that the electrical vector of the heart is pointing from base (negative) towards the apex (positive) of the heart during both depolarization and repolarization.

Electrical Axis of the Heart. The heart has an electrical axis during ventricular depolarization and repolarization—negative in the base (near atria) and positive in the apex (tip). Recognizing the orientation of the electrical axis of the heart is helpful for understanding the positioning of the twelve ECG leads and the polarity of electrocardiogram recorded by the twelve ECG leads. When an ECG lead is positioned in the same direction as the electrical axis of the heart, for example, Lead II, the QRS complex and T wave are recorded as positive voltages (Fig. 6.6B). When an ECG lead is positioned opposite to the direction of the electrical axis, for example, Lead aV_R, the QRS and T wave are recorded as negative voltages.

To understand the electrical axis of the heart during ventricular depolarization and repolarization, it is helpful to consider the ventricle as one giant cell with a base and an apex (Fig. 6.7A), based on the rationale that cardiac muscle cells are connected electrically by gap junctions. The base of the ventricle is the area near the atria. The apex of the ventricle is the tip of the ventricle, which is pointing downward. In the example shown in Fig. 6.7A, the positive terminal of an electrocardiograph is placed on the ventricular apex and the negative terminal of an electrocardiograph is placed on the ventricular base. The **top panel in** Fig. 6.7A shows the ventricle in the relaxed state, when all cardiac muscle cells are uniformly repolarized (negative inside the cell relative to the outside), and the electrocardiograph records zero voltage. The **middle panel in** Fig. 6.7A shows the beginning of ventricular depolarization, when cardiac muscle cells at the ventricular base become depolarized (positive inside the cell relative to the outside), whereas cardiac muscle cells at the ventricular apex remain in the repolarized state (negative inside the cell relative to the outside), and the electrocardiograph records a positive voltage. The **bottom panel in** Fig. 6.7A shows the completion of ventricular depolarization, when the ventricle is uniformly depolarized (positive inside the cell relative to the outside), and the ECG records zero voltage.

Fig. 6.7B shows the electrical axis of the heart during ventricular repolarization. The **top panel in** Fig. 6.7B shows the ventricle at the completion of depolarization, when the ventricle is uniformly depolarized (positive inside the cell relative to the outside). The **middle panel** in Fig. 6.7B shows that the apex of the ventricle becomes repolarized (negative inside the cell relative to the outside), whereas the base of the ventricle remains in the depolarized state (positive inside the cell relative to the outside), and the ECG records a positive voltage. It is noteworthy that the apex of the ventricle is the last part to become depolarized but the first part to become repolarized, possibly because the ventricular apex experiences the lowest mechanical stress in the heart. The **lower panel in** Fig. 6.7B shows the completion of repolarization, when the ventricle becomes uniformly repolarized (negative inside the cell relative to the outside), and the electrocardiograph records zero voltage.

In summary, Fig. 6.7 shows that the electrical axis of the ventricle during both the depolarization and repolarization waves (QRS and T wave in an electrocardiogram) is positive at the ventricular apex and negative at the ventricular base. The electrical axis of the ventricle provides the rationale for the attachment of the positive and negative terminals of standard ECG leads I, II, and III to the limbs. Recognizing the electrical axis of the ventricle is also helpful for understanding the different polarity of electrocardiograms recorded by the twelve ECG leads.

Twelve-Lead Electrocardiogram. The heart is a three-dimensional organ. A standard twelve-lead ECG is designed to measure electrical activity of the heart from twelve directions by attaching ECG electrodes to the body surface in twelve configurations.

Standard Leads (I, II, and III). As shown in Fig. 6.8A, the general configuration for the three standard leads (I, II, and III) is that the positive and negative terminals are attached to different limbs. For Lead I, the positive terminal is attached to the left arm, and the negative terminal is attached to the right arm. For Lead II, the positive terminal is attached to the left leg, and the negative terminal is attached to the right arm. For Lead III, the positive terminal is attached the left leg, and the negative terminal is attached to the left arm.

Understanding the electrical axis of the ventricle is helpful for understanding the attachment of the three standard leads to the limbs. The electrical axis of the heart is positive at the ventricular tip and negative at the ventricular base—that is, negative in the upper right part of the body and positive in the lower left part of the body. As shown in Fig. 6.8A, the three standard leads are oriented along the direction of the electrical axis of the heart, and QRS complex and T wave recorded by the three standard ECG leads typically exhibit positive voltages. Among the three standard leads, Lead II ECG typically exhibits the highest voltage, because the directionality of Lead II is almost identical to the electrical axis of the heart.

Augmented Leads. As shown in Fig. 6.8B, the general configuration for augmented leads is that the positive terminal is attached to one limb, and the negative terminal is attached to two other limbs. For Lead aV_R, the positive terminal is attached to the right arm, and the negative terminal is attached to the left arm and left leg. For Lead aV_L, the positive terminal is attached to the left arm, and the negative terminal is attached to the right arm and left leg. For Lead aV_F, the positive terminal is attached to the left leg, and the negative terminal is attached to the right and left arms. Augmented leads are oriented along the middle between two standard leads. For example, Lead aVR is oriented along the middle between leads I and II. Lead aVL is oriented along the middle between leads I and III. Lead aV_L is oriented along the middle between leads II and III.

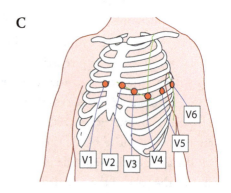

Fig. 6.8. The 12 leads of electrocardiogram. **A.** Standard leads I, II, and III. **B.** Augmented leads aVR, aVL, and aVF. **C.** Precordial leads V1 to V6.

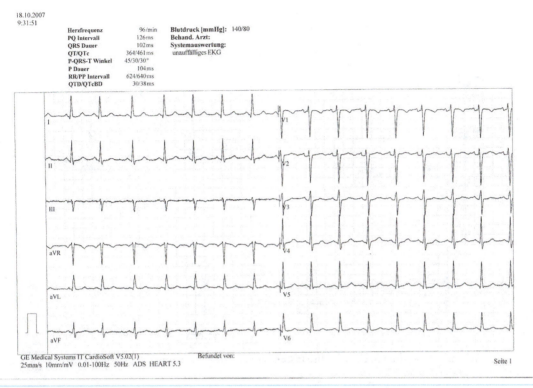

RUHE-EKG / 2 * 5s

18.10.2007
9:31:51

Herzfrequenz	96/min	Blutdruck [mmHg]:	140/80
PQ Intervall	126ms	Behand. Arzt:	
QRS Dauer	102ms	Systemauswertung:	
QT/QTc	364/461ms	unauffälliges EKG	
P-QRS-T Winkel	45/30/30°		
P Dauer	104ms		
RR/PP Intervall	624/640ms		
QTD/QTcBD	30/38ms		

GE Medical Systems IT CardioSoft V5.02(1)   Befundet von:
25mm/s 10mm/mV 0.01-100Hz 50Hz ADS HEART 5.3   Seite 1

Fig. 6.9. A normal 12-lead electrocardiogram.

As shown in Fig. 6.9, QRS complex and T wave recorded by Lead aVR are typically inverted (negative voltage), because the positive terminal of Lead aVR is placed near the negative end of the electrical axis of the heart, and the negative terminal of Lead aVR is placed near the positive end of the electrical axis. QRS complex and T wave recorded by leads aV_L and aV_F typically exhibit positive voltages, because the directionality of these two leads is consistent with the electrical axis of the ventricle.

Precordial Leads. As shown in Fig. 6.8C, the configuration for the six precordial leads is that the positive terminal is attached to one point on the chest wall, and the negative terminal is attached to the right arm, left arm, and left leg together. Precordial leads may be considered as oriented from the center of the heart toward one point on the surface on the chest wall. As shown in Fig. 6.8C, positive terminals for leads V1 and V2 are placed on the two sides of the mid-sternum. Positive terminals for leads V3 to V6 are placed along the side of the left chest wall from the front toward the back of the body.

As shown in Fig. 6.9, QRS complex and T wave recorded by leads V1 and V2 are often inverted, because the positive terminals for these two precordial leads are placed near the ventricular base—the negative end of the electrical axis of the heart. QRS complex and T wave recorded by leads V3 to V6

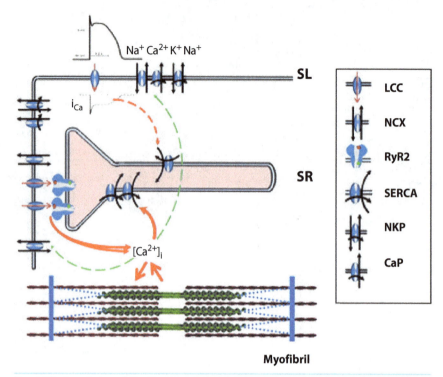

Fig. 6.10—Regulation of Cardiac Muscle Contraction. (From: ter Keurs HEDJ Am J Physiol Heart Circ Physiol 302: H39, 2012).

are typically upright (positive voltages), because the positive terminals for these two precordial leads are placed near the positive end of the electrical axis of the heart.

CARDIAC MUSCLE

Cardiac muscle cell is a striated muscle cell. Similar to skeletal muscle contraction, cardiac muscle contraction is regulated by the Ca^{2+}-troponin-tropomyosin system. Unlike skeletal muscle cells, the increase in intracellular $[Ca^{2+}]$ in cardiac muscle cells in response to action potentials is mediated by two mechanisms—Ca^{2+} influx across the cell membrane and intracellular Ca^{2+} release from the sarcoplasmic reticulum. As shown in Fig. 6.10, membrane depolarization during the plateau depolarization of a cardiac action potential stimulates Ca^{2+} influx into cardiac muscle cells through voltage-gated L-type calcium channels (LCC) on the cell membrane. The initial increase in intracellular $[Ca^{2+}]$ then triggers Ca^{2+} release from the sarcoplasmic reticulum via ryanodine channels (RyR2)—a mechanism known as Ca^{2+}-induced Ca^{2+} release. At the termination of an action potential, cardiac muscle relaxation is induced by two mechanisms. First, intracellular Ca^{2+} is pumped back to the sarcoplasmic reticulum by Ca^{2+}-ATPase (SERCA) on the sarcoplasmic reticular membrane. Second, intracellular Ca^{2+} is extruded out of the cell to the interstitium by Na^+/Ca^{2+} exchanger (NCX) and Ca^{2+}-ATPase (CaP) on the cell membrane. Intracellular Na^+ is pumped out of cardiac muscle cells by Na^+-K^+-ATPase (NKP).

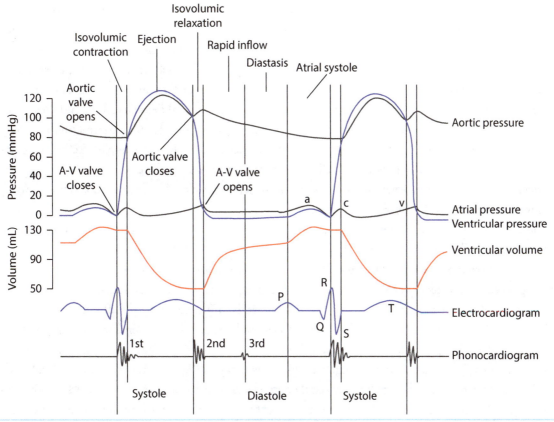

Fig. 6.11. Electrical and mechanical events during a cardiac cycle.

The binding of Ca^{2+} to troponin causes conformational change in the troponin-tropomyosin complex and removes the inhibitory effect of the complex on actin filament, thereby allowing cyclic interactions between actin with myosin crossbridges for muscle contraction. Amino acid sequences of contractile proteins in cardiac and skeletal muscle cells are similar but not identical. For example, cardiac muscle and skeletal muscle troponins can be detected differentially using specific antibodies in the diagnosis of heart attack patients. The presence of cardiac troponin in the plasma of heart attack patients is an indicator of significant damage to cardiac muscle cells and a poor prognosis for survival.

CARDIAC CYCLE

Cardiac Cycle refers to the cycles of atrial and ventricular contractions that enable the heart to circulate blood from the venous system to the arterial system. Right and left cardiac cycles are temporally synchronized, because right and left ventricular cardiac muscle cells are electrically synchronously by the fast conductive system. The discussion of cardiac cycle often focuses on the left ventricle, because the left ventricle does most of the work in pumping cardiac output against the high resistance in the systemic circulation. In comparison, the right ventricle pumps cardiac output against a relatively low resistance in the pulmonary circulation. The two atria function largely as a conduit for venous return to the heart. Atrial contractions contribute to only 10 to 20 percent of the filling of ventricles.

Fig. 6.11 shows the electrocardiogram, left ventricular volume, left ventricular pressure, aortic pressure, and atrial pressure during a cardiac cycle. The electrocardiogram is an important marker of events during a cardiac cycle. For example, the P wave marks atrial depolarization and atrial contraction; the QRS complex marks ventricular depolarization and ventricular contraction (systole); and T wave marks ventricular repolarization and ventricular relaxation (diastole). Left ventricular volume and left ventricular pressure together define the four phases of cardiac cycle. Aortic pressure is determined by periodic ejections of blood from the left ventricle into the arterial system and continuous blood flow from the arterial system to the venous system via organ systems. Atrial pressure is determined by the continuous inflow of blood from the venous system to the atria, and periodic outflow of blood from the atria into the ventricles during ventricular filling.

Left Ventricular Volume During a Cardiac Cycle. The left ventricular volume is a good variable to begin the discussion of cardiac cycle, because the four phases of cardiac cycle are clearly defined by the time course of left ventricular volume (Fig. 6.11, **red line**). The onset of ventricular depolarization, as marked by the QRS complex in an electrocardiogram, is a convenient starting point for examining the changes in left ventricular volume during a cardiac cycle. At the time immediately before the QRS complex, the left ventricular volume reaches its highest level (end-diastolic volume), at the end of ventricular filling after left atrial contraction. The mitral valve remains open at the end of ventricular filling, because left atrial and left ventricular pressures are similar. The aortic valve remains closed, because left ventricular pressure is significantly lower than aortic pressure.

Isovolumic Contraction Phase. Onset of the QRS complex marks the beginning of left ventricular contraction and increase in left ventricular pressure, which will first cause the closing of the mitral valve and then opening of the aortic valve for ejection of blood into the circulation. As shown in Fig. 6.11 (**red line**), at the beginning of left ventricular contraction, there is a brief period between the closing of the mitral valve and opening of the aortic valve—known as the isovolumic contraction phase—when ventricular volume remains constant but left ventricular pressure is rising. During the isovolumic contraction phase, both mitral and aortic valves are closed, because left ventricular pressure is higher than left atrial pressure but lower than the aortic pressure.

Ejection Phase. Shortly after isovolumic contraction, the left ventricle enters the ejection phase, when left ventricular pressure exceeds the aortic pressure to cause the opening of the aortic valve and ejection of blood (stroke volume) into the circulation. As shown in Fig. 6.11 (**red line**), the left ventricular volume decreases during the ejection phase and reaches its lowest level (end-systolic volume) at the end of ejection phase. Blood volume ejected into the circulation during one cardiac cycle is defined as **stroke volume**, which is the difference between **end-diastolic volume** and **end-systolic volume**, as shown in the following equation:

$$\text{Stroke Volume} = \text{End-Diastolic Volume} - \text{End-Systolic Volume}$$

End-systolic volume is typically nonzero, because the heart does not completely empty its content into the circulation during the ejection phase. The fraction of end-diastolic volume that is ejected as stroke volume into the circulation during one cardiac cycle is defined as ejection fraction, as shown in the following equation:

$$\text{Ejection Fraction} = \text{Stroke Volume}/\text{End-Diastolic Volume}$$

Ejection fraction is a measure of cardiac contractility—contractile strength of a heart. Ejection fraction in a healthy heart is relatively high (> 50%), whereas ejection fraction in a failing heart is relatively low (< 30%).

Isovolumic Relaxation Phase. Onset of the T wave in an electrocardiogram marks the end of ejection phase and the beginning of ventricular repolarization, ventricular diastole (relaxation), and closing of the aortic valve. As shown in Fig. 6.11, there is a brief period between the closing of the aortic valve and opening of the mitral valve—known as isovolumic relaxation—when the left ventricular volume remains constant but the left ventricular pressure is falling. During the isovolumic relaxation phase, both mitral and aortic valves are closed, because left ventricular pressure is lower than the aortic pressure but higher than the left atrial pressure.

Filling Phase. Shortly after isovolumic relaxation, the left ventricle enters the passive filling phase, when left ventricular pressure falls below the left atrial pressure, causing the mitral valve to open and blood from the left atrium to enter the left ventricle. During the passive filling phase, ventricular relaxation drives blood flow from the atrium into the ventricle. As shown in Fig. 6.11 (**red line**), passive filling contributes to the majority (80 to 90 percent) of left ventricular filling. Sometime later, the onset of the P wave in an electrocardiogram marks the beginning of atrial depolarization, atrial contraction, and active filling phase, during which atrial contraction drives blood flow from the atrium to the ventricle. At the end of the active filling phase, the left ventricular volume reaches its highest level (end-diastolic volume).

The temporal sequence of the four phases of cardiac cycle is shown as follows:

$$\text{Isovolumic Contraction} \rightarrow \text{Ejection} \rightarrow \text{Isovolumic Relaxation} \rightarrow \text{Filling}$$

The four phases of cardiac cycle can be grouped into systole (contraction) and diastole (relaxation).

$$\text{Systole} = \text{Isovolumic Contraction} + \text{Ejection}$$
$$\text{Diastole} = \text{Isovolumic Relaxation} + \text{Filling}$$

Left Ventricular Pressure is the major driver of both ventricular ejection and ventricular filling. As shown in Fig. 6.11 (**blue line**), the left ventricular pressure fluctuates over a large range—for example, diastolic ventricular pressure is typically near zero, whereas ventricular systolic pressure is relatively high, such as 120 mmHg. A low ventricular diastolic pressure is necessary for ventricular filling. A high ventricular systolic pressure is necessary for overcoming the aortic pressure during ventricular ejection.

As shown in Fig. 6.11 (**blue line**), the QRS complex of an electrocardiogram marks ventricular depolarization and the beginning of ventricular systole (contraction). At the time immediately before the QRS complex, left ventricular pressure is at the lowest (diastolic) level, because the left ventricle is in the relaxed state. The onset of the QRS complex marks the beginning of ventricular depolarization, ventricular contraction, and rise in left ventricular pressure. For a short period, the left ventricle enters the **isovolumic contraction phase**, when left ventricular pressure is above the left atrial pressure but below the aortic pressure, during which the mitral and aortic valves are both closed and the left ventricular volume remains constant, while the left ventricular pressure is rising. Shortly after, the left ventricle enters the ejection phase, when left ventricular pressure rises above the aortic pressure, causing the aortic valve to open, and the left ventricle to eject blood into the aorta. The left ventricular pressure continues to rise during the ejection phase, reaching its highest (systolic) level in the middle of the ejection phase before falling to a lower level during the later part of the ejection phase due to the decrease in ventricular volume.

Onset of the T wave of an electrocardiogram marks the beginning of ventricular repolarization, ventricular relaxation (diastole), and a precipitous fall in left ventricular pressure. When left ventricular pressure falls below the aortic pressure, but is still above the left atrial pressure, both aortic and mitral valves close, the left ventricle enters the **isovolumic relaxation phase**. Shortly after, when left ventricular pressure falls below the left atrial pressure, causing the mitral valve to open and atrial blood to flow into the left ventricle, the left ventricle enters the **passive filling phase**. Sometime later, the onset of the P wave of an electrocardiogram marks the beginning of atrial depolarization, atrial contraction, and **active filling phase**, when atrial contraction drives blood flow from the left atrium into the left ventricle. During atrial contraction, pressure from the atrium is transmitted to the ventricle via the open mitral valve, causing a transient increase in left ventricular pressure. At the completion of the filling phase, the pressure in the completely relaxed left ventricle decreases to its lowest (diastolic) level, which is typically near zero.

Aortic Pressure. The aorta receives blood periodically from the left ventricle during the ejection phase and continuously propels blood to the venous system via organ systems. It is noteworthy that the left ventricle ejects blood into the aorta only during the ejection phase, which occupies approximately 40 percent of cardiac cycle time. Ventricular output to the aorta is zero during the other three phases of cardiac cycle, when the aortic valve remains closed. Despite the lack of ventricular output, blood flow through the systemic circulation does not fall to zero, because the arterio-venous pressure difference continues to drive blood flow through the systemic circulation.

As shown in Fig. 6.11, the aortic pressure fluctuates between a lowest (diastolic) level to a highest (systolic) level during a cardiac cycle. For example, a typical value of aortic pressure is 120/80 mmHg—that is, 80 mmHg diastolic pressure and 120 mmHg systolic pressure. As shown in Fig. 6.11, after the onset of the QRS wave—marking ventricular depolarization and ventricular contraction—there is a time delay between the rise in left ventricular pressure and rise in aortic pressure. The delay represents the duration of isovolumic contraction phase, when the left ventricular pressure is rising toward the level above the aortic pressure for ventricular ejection. After the opening of the aortic valve, the left ventricle ejects blood into the aorta, causing the aortic pressure to rise to the highest systolic level in the middle of the ventricular ejection phase. The aortic pressure is almost identical to left ventricular pressure during the ejection phase, because the vascular resistance of the large aorta is relatively low.

The T wave of an electrocardiogram marks the beginning of ventricular repolarization, ventricular relaxation, and a precipitous fall in left ventricular pressure, which causes the closing of the aortic valve and dissociation between left ventricular pressure and aortic pressure. The left ventricular pressure continues to fall rapidly toward near zero for filling by the atrium. In the absence of ventricular output, the aortic pressure falls gradually toward the diastolic pressure—due to the outflow of blood from the aorta to the venous system via the organ systems—until the next ventricular ejection.

Left Atrial Pressure. Atrial pressures are relatively low (typically less than 5 mmHg) during a cardiac cycle. The left atrium receives blood from pulmonary veins. The small pressure fluctuations

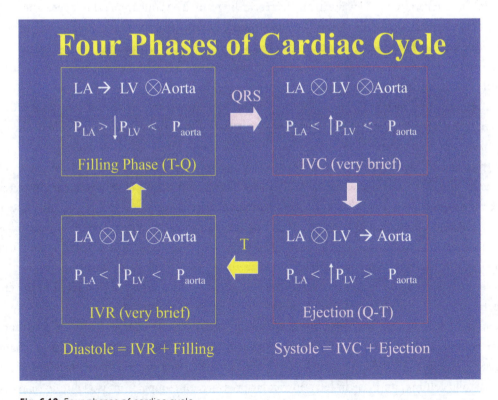

Fig. 6.12. Four phases of cardiac cycle.

in the left atrium reflect the venous return to the atrium, contraction of the atrium, and contraction of the left ventricle.

Fig. 6.11 shows the time course of the left atrial pressure during a cardiac cycle. The QRS wave marks the beginning of ventricular depolarization, ventricular contraction, and a transient increase in the left atrial pressure (**c wave**), which is caused by the vibration of the mitral valve during ventricular contraction. The left atrial pressure increases during ventricular ejection due to the accumulation of pulmonary venous return in the left atrium. The T wave of an electrocardiogram marks the beginning of ventricular repolarization and ventricular relaxation. When the left ventricular pressure falls below the left atrial pressure, the mitral valve opens, causing blood flow from the left atrium into the relaxed left ventricle and a decrease in left atrial pressure. The **v wave** of the left atrial pressure reflects the transition from the accumulation of venous return in the left atrium to outflow of blood from the left atrium to the left ventricle. Sometime after the opening of the mitral valve, the P wave of an electrocardiogram marks atrial depolarization and atrial contraction, which causes a transient increase in left atrial pressure (**a wave**).

Phonocardiogram. Heart sounds originate from closing of cardiac valves in pairs: lub-dub, lub-dub … lub-dub. The first heart sound—lub (S1)—originates from the synchronized closing of tricuspid and mitral valves at the beginning of ventricular contraction (systole), as marked by the QRS wave in an electrocardiogram (Fig. 6.11). The second heart sound—dub (S2)—originates from the synchronized closing of the aortic and pulmonary valves at the beginning of ventricular relaxation (diastole), as marked by the T wave in an electrocardiogram. The time interval between the first and second heart sounds (lub-dub interval) corresponds to systolic time. The time interval between a consecutive pair of heart sounds (dub-lub interval) corresponds to diastolic time. Systolic time is typically shorter than diastolic time. The third heart sound (S3) that can occur in early diastole is relatively rare in normal subjects, and is considered as an indicator of cardiac failure.

Four Phases of Cardiac Cycle. Fig. 6.12 summarizes the status of the left atrial pressure, left ventricular pressure, and aortic pressure and the status of the mitral and aortic valves during the four phases of cardiac cycle. The filling phase is characterized by a left ventricular pressure that is lower than the left atrial pressure and lower than the aortic pressure, causing the opening of the mitral valve and closing of the aortic valve. The isovolumic contraction (IVC) phase is characterized by a left ventricular pressure that is higher than the left atrial pressure but lower than the aortic pressure, causing the closing of both mitral and aortic valves. The ejection phase is characterized by a left ventricular pressure that is higher than both left atrial pressure and aortic pressure, causing closing of the mitral valve and opening of the aortic valve. The isovolumic relaxation (IVR) phase is characterized by a left ventricular pressure that is higher than the left atrial pressure but lower than the aortic pressure, causing the closing of both aortic and mitral valves.

Isovolumic contraction and isovolumic relaxation phases are similar in having a constant ventricular volume, but different in the size of ventricular volume and the directionality of change in ventricular pressure. Isovolumic contraction phase is associated with a high ventricular volume (end-diastolic volume) and rise in left ventricular pressure. In comparison, isovolumic relaxation is associated with a low ventricular volume (end-systolic volume) and a fall in left ventricular pressure.

REGULATION OF CARDIAC OUTPUT

The primary function of the heart is to pump blood from the venous system into the arterial system. Cardiac output—the volume of blood that is pumped by the heart into the circulation per min—in a normal person is approximately 5,000 ml/min at rest and can increase to 12,500 ml/min during exercise. Cardiac output is determined by the heart rate and stroke volume, as shown in the following equation:

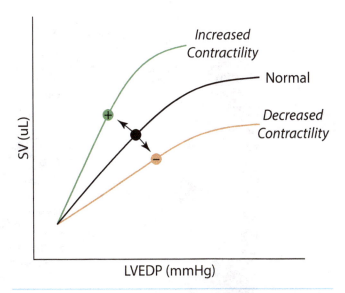

Fig. 6.13—Frank-Starling Curves for Describing the Dependence of Stroke Volume (SV) on Left Ventricular End-diastolic Pressure (LVEDP). This diagram shows a family of three Frank-Starling curves for different levels of cardiac contractility. An increase in contractility shifts the curve to the left and higher position, whereas a decrease in contractility shifts the curve to the right and lower position. At a given LVEDP, stroke volume is higher at higher contractility and lower at lower contractility.

Cardiac Output = Heart Rate × Stroke Volume

Heart rate is the number of heart beats per minute. Stroke volume is the volume (mls) of blood ejected by the heart into the circulation in each cardiac cycle. One determinant of cardiac output is heart rate. The other three determinants of cardiac output—preload, afterload, and cardiac contractility—regulate stroke volume.

Heart Rate. As addressed previously, heart rate is regulated by the sinoatrial node under modulation by the autonomic nervous system. Parasympathetic (vagal) stimulation of sinoatrial node decreases heart rate, whereas sympathetic stimulation of the sinoatrial node increases heart rate.

Preload is the extent of ventricular filling—as measured by left ventricular end-diastolic volume or pressure—prior to ventricular contraction. As shown in Fig. 6.13, an increase in preload (left ventricular end-diastolic pressure) leads to an increase in stroke volume—a relation known as Frank-Starling curve for the heart. Preload is determined mostly by blood volume and venous compliance. For example, severe hemorrhage and dehydration can lead to a decrease in blood volume and decrease in preload, resulting in a decrease in stroke volume and cardiac output.

Afterload is the load against which the heart ejects its stroke volume. A simple measure of afterload is mean arterial pressure, which the left ventricular pressure must exceed to open the aortic valve for ejection. In a normal heart, moderate increase in afterload (mean arterial pressure) does not significantly change stroke volume due to the presence of compensatory mechanisms such as cardiac contractility modulation. In a failing heart, an increase in afterload (mean arterial pressure) can result in a decrease in stroke volume due to the exhaustion of compensatory mechanisms. In

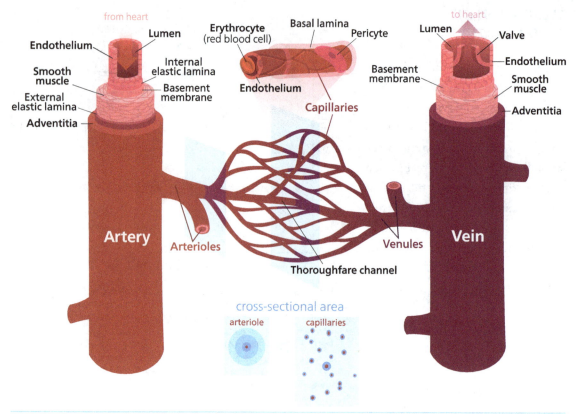

Fig. 6.14. The vascular system.

both normal and failing hearts, an increase in afterload (mean arterial pressure) leads to an increase in the workload and oxygen consumption by the heart, which could result in cardiac failure.

Among the four determinants of cardiac output, heart rate and cardiac contractility are properties of the heart alone. In comparison, preload and afterload are dependent on both the heart and the vascular system. For example, preload is dependent on venous return to the heart from the vascular system, and afterload (mean arterial pressure) is dependent on total vascular resistance of the circulation.

Cardiac Contractility is defined conceptually as the contractile strength of the heart. Cardiac contractility can be measured in terms of the Frank-Starling curve. Fig. 6.13 shows a family of three Frank-Starling curves for different levels of cardiac contractility. As shown in Fig. 6.13, an increase in cardiac contractility shifts the curve to the left and upward direction, and a decrease in contractility shifts the curve to the right and downward direction. Another measure of cardiac contractility is the ejection fraction—stroke volume/end-diastolic volume. At a given end-diastolic volume, a stronger heart with higher contractility ejects a larger stroke volume, and a weaker heart with lower contractility ejects a smaller stroke volume. Sympathetic stimulation of the heart increases cardiac contractility directly by activating β-adrenergic receptors on cardiac muscle cells with the sympathetic neurotransmitter—norepinephrine. Sympathetic stimulation of the adrenal medulla induces the

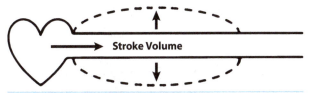

Fig. 6.15—The Arterial System Functions as a Pressure Reservoir. The arterial system expands during ventricular ejection. The arterial system recoils after the closing of aortic valve during ventricular relaxation to sustain arterial pressure for driving blood flow to organ systems.

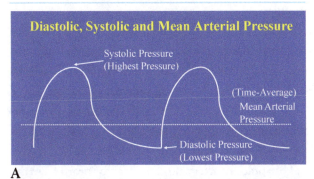

Diastolic, Systolic and Mean Arterial Pressure

Systolic Pressure (Highest Pressure)

(Time-Average) Mean Arterial Pressure

Diastolic Pressure (Lowest Pressure)

A

B

Fig. 6.16—A. Diastolic, systolic and mean arterial pressure. **B.** Measurement of blood pressure.

release of epinephrine, which increases cardiac contractility by stimulating β-adrenergic receptors on cardiac muscle cells. Parasympathetic stimulation of the heart decreases heart rate, but does not directly alter the contractility of cardiac muscle cells.

Heart rate is an indirect determinant of cardiac contractility, because intracellular $[Ca^{2+}]$ in cardiac muscle cells increases with frequency of stimulation. By modulating heart rate, parasympathetic stimulation indirectly decreases cardiac contractility, whereas sympathetic stimulation directly increases cardiac contractility.

THE VASCULAR SYSTEM

Fig. 6.14 shows the architecture of the vascular system in a typical organ—consisting of arteries, arterioles, capillaries, venules, and veins. Kidney and liver are exceptions to this structure in having two arterioles and two capillary beds. Renal and hepatic circulations will be examined in detail in the chapters on renal and gastrointestinal physiology.

In most organs other than the lungs, an artery carries oxygenated blood to the organ, and a vein carries deoxygenated blood away from the organ. Within an organ, the artery branches into many arterioles, and each arteriole branches into many capillaries, where diffusional exchange occurs between the blood and surrounding cells. Capillaries then converge to form venules, and the venules converge to form a vein. All segments of the circulation have a lumen that is covered by a single layer of endothelial cells, which performs multiple functions—for example, providing a nonclotting surface for blood flow and releasing nitric oxide for vasodilation. Each vascular segment has a unique structure and performs a specific function. For example, arteries have a relatively thick and stiff wall for withstanding the relatively high arterial pressure and storing mechanical energy during ventricular ejection. Arterioles contain a large amount of vascular smooth muscle for controlling the arteriolar diameter in the regulation of blood flow. Capillaries are made up of a single layer of endothelial cells for rapid diffusional transport between blood and the surrounding cells. Venules and veins have relatively thin and compliant walls for holding majority of the blood volume in the circulation.

Arteries. Arteries function as a pressure reservoir for maintaining continuous blood flow to organs, despite intermittent ventricular ejections of stroke volume to the circulation. As shown in Fig. 6.15, the arterial system functions as a pressure reservoir when the left ventricle pumps blood (stroke volume) into the arterial system during ventricular ejection, causing expansion of the arterial wall. The stored mechanical pressure in the arterial system is used to drive continuous blood flow despite intermittent ejection of blood into the circulation.

Fig. 6.16A shows the time course of arterial pressure. Arterial pressure is expressed in the systolic/diastolic format, where **systolic pressure** represents the highest pressure and **diastolic pressure** represents the lowest pressure. For example, normal arterial pressure is 120/80 mmHg. Systolic arterial pressure is reached in the middle of ventricular ejection. Diastolic arterial pressure is reached at the end of isovolumic contraction, immediately before the ejection phase. Fig. 6.16B shows the measurement of arterial pressure by placing a pressure cuff around the upper arm and listening to the brachial artery below the pressure cuff using a stethoscope. When the cuff pressure is increased to a level higher than systolic arterial pressure, causing occlusion of the brachial artery, no sound can be heard from the artery distal from the pressure cuff. When the cuff pressure is decreased to a level just below systolic arterial pressure, causing partial opening of the brachial artery, sound is emitted from turbulent blood flow in the artery. The cuff pressure at which the first sound appears is the systolic arterial pressure. When the cuff pressure is decreased gradually, sound continues to be emitted from the artery until the cuff pressure falls to the level below diastolic arterial pressure, causing full opening of the brachial artery. The cuff pressure at which the sound disappears marks the diastolic arterial pressure.

The difference between systolic pressure and diastolic pressure is **pulse pressure**, as shown in the following equation:

$$\text{Pulse Pressure} = \text{Systolic Pressure} - \text{Diastolic Pressure}$$

For example, for an arterial pressure of 120/80 mmHg, pulse pressure is 40 mmHg. Pulse pressure represents the change in arterial pressure in response to ejection of stroke volume into the arterial system. Accordingly, stroke volume and arterial compliance are major determinants of pulse pressure. For example, a decrease in stroke volume as a result of severe hemorrhage can lead to a decrease in pulse pressure. Stiffening of the arterial wall as a result of atherosclerosis can lead to an increase in pulse pressure.

Mean arterial pressure—time-average of arterial pressure—is a useful indicator of blood pressure homeostasis. In principle, mean arterial pressure is calculated by averaging values of arterial blood pressure measured at regular time intervals during a cardiac cycle. An approximate estimate of mean arterial pressure can be calculated using the following equation:

$$\text{Mean Arterial Pressure} = \text{Diastolic Pressure} + \text{Pulse Pressure}/3$$

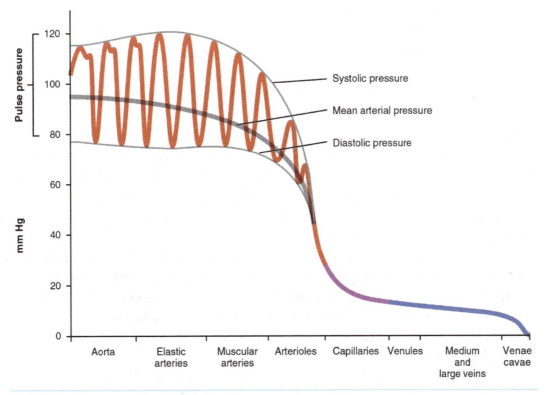

Fig. 6.17. Blood pressure profile along the circulation.

For example, for an arterial pressure of 120/80 mmHg, the estimated mean arterial pressure is 80 + (120-80)/3 = 93 mmHg. Mean arterial pressure is closer to diastolic pressure than systolic pressure, because ventricular diastole occupies majority (60 percent) of cardiac cycle time.

Arterioles. Arterioles Function as Resistance Vessels for the regulation of blood flow to organs. The function of arterioles as the major site of vascular resistance is reflected by the profile of blood pressure drop through each segment of the vascular system. It is noteworthy that, for a given organ, each segment of the vascular system is perfused by the same blood flow, but the pressure drop through each segment is different, depending on the vascular resistance in the segment, as shown in the following equation:

$$\textbf{Vascular Resistance} = (\textbf{Pressure}_{in} - \textbf{Pressure}_{out}) / \textbf{Blood Flow}$$

The above equation predicts that the pressure drop through each segment is proportional to the vascular resistance in the segment. As shown in Fig. 6.17, the arterioles segment has the largest pressure drop, because it has the highest vascular resistance among all vascular segments.

Paradoxically, capillaries are the smallest blood vessels in the vascular system, but total vascular capillary resistance is smaller than total arteriolar resistance. This paradox may be understood by analyzing the relation between the vascular resistance of a single blood vessel and total vascular resistance of many blood vessels connected in parallel. The resistance of a single blood vessel can be calculated using **Poiseuille's law**, as shown in the following equation:

$$\textbf{Resistance} = \textbf{8} \times \textbf{Length} \times \textbf{viscosity} \ / \ (\pi \times \textbf{radius}^4)$$

This equation predicts that vascular resistance of a single blood vessel is linearly proportional to vessel length and viscosity, and inversely proportional to the fourth power of radius. Among the three variables in the above equation, radius is the most important determinant of vascular resistance, for example, a twofold increase in radius will lead to sixteenfold decrease in resistance.

Poiseuille's law is useful for explaining the low vascular resistances in large blood vessels (aorta and large arteries) and high vascular resistances in small blood vessels (arterioles and capillaries). Poiseuille's law predicts that single arteriolar resistance is lower than single capillary radius, because arteriolar radius is larger than capillary radius. The paradox that total arteriolar resistance is higher than total capillary resistance can be explained by considering the number of capillaries and arterioles. Total vascular resistance (R_{total}) of a network of multiple blood vessels arranged in parallel can be calculated using the following equation, where R_1, R_2,...,R_n represent the single vascular resistances of individual blood vessels:

$$1/R_{total} = 1/R_1 + 1/R_2 + ... + 1/R_n$$

For a network of parallel blood vessels with identical single vascular resistances, the above general equation can be simplified to the following equation, where N represents the number of blood vessels arranged in parallel:

$$1/R_{total} = N/R_{single}$$

or

$$R_{total} = R_{single}/N$$

Total arteriolar resistance is higher than total capillary resistance, despite the fact that single arteriolar arteriolar resistance is lower than single capillary resistance, because the number of arterioles is much lower than the number of capillaries.

As shown in Fig. 6.17, the total arterial-venous pressure difference is the sum of individual pressure drops in each segment of the vascular system. Since pressure drop represents vascular resistance in each segment, total vascular resistance in an organ represents the sum of vascular resistances in the individual segments, as shown in the following equation:

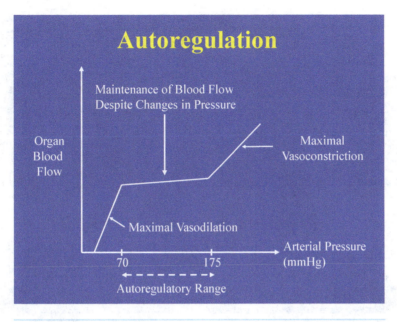

Fig. 6.18. Autoregulation of blood flow.

$$\text{Total Vascular Resistance in an Organ} = R_{\text{artery}} + R_{\text{arteriole}} + R_{\text{capillary}} + R_{\text{venule}} + R_{\text{vein}}$$

As shown in the above equation, total vascular resistance in an organ can be regulated by controlling the resistance of arterioles by vasoconstriction and vasodilation. Regulation of total vascular resistance is important for the regulation of blood flow to an organ, as shown by the following equation:

$$\text{Organ Blood Flow} = (\text{Arterial Pressure} - \text{Venous Pressure}) / \text{Vascular Resistance}$$

Neural Control of Blood Flow. In kidneys and the gastrointestinal tract, sympathetic stimulation of arteriolar constriction plays a major role in the regulation of organ blood flow. In many organs—in particular, the heart and brain—local mechanism dominates the regulation of organ blood flow. For example, during intense exercise, coronary blood flow increases dramatically to support metabolism of the heart, despite the high level of sympathetic stimulation.

Local Control of Blood Flow. Autoregulation is the maintenance of relatively constant organ blood flow by local mechanisms over a range of arterial pressure, as illustrated in Fig. 6.18. The venous pressure is kept at zero in this experiment. As shown in Fig. 6.18, organ blood flow begins at a minimum pressure—critical closing pressure—a pressure that is necessary for overcoming the collapsibility of the vascular system within an organ. Blood flow then increases linearly with arterial pressure between critical closing pressure and 70 mmHg, reflecting a constant minimal vascular resistance, when arterioles are maximally dilated. As shown in Fig. 6.18, between 70 to 175 mmHg (autoregulatory range), organ blood flow is maintained at a relatively constant level independent of arterial

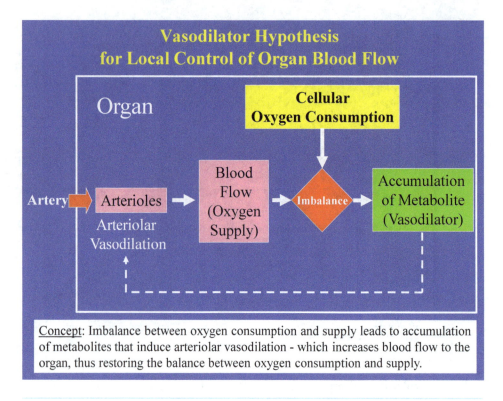

Fig. 6.19. Vasodilator hypothesis for local control of blood flow.

pressure. The maintenance of relatively constant organ blood flow within the autoregulatory range can be explained by proportional increases in vascular resistance with arterial pressure, as predicted by the following equation:

Organ Blood Flow = (Arterial Pressure – Venous Pressure) / Vascular Resistance

Beyond the autoregulatory range, organ blood flow increases linearly with arterial pressure, reflecting a constant maximal vascular resistance, when arterioles are maximally constricted.

The mechanism of blood flow autoregulation is addressed in the following section.

Vasodilator Hypothesis for Explaining Local Regulation of Blood Flow. The vasodilator hypothesis explains the maintenance of constant blood flow despite a decrease in arterial pressure within the autoregulatory range by proposing that a decrease in arterial pressure will initially lead to insufficient blood flow and oxygen for cellular metabolism, causing an increase in the release of vasodilator by cells. The accumulation of vasodilator will then lead to arteriolar dilation and an increase in blood flow until oxygen supply matches oxygen demand. A major assumption in the vasodilator hypothesis is that a vasodilator—a product of anaerobic metabolism—is released by cells when the supply of oxygen is insufficient to meet the demand of cellular metabolism, as shown in Fig. 6.19. Conversely, the vasodilator hypothesis explains the maintenance of blood flow

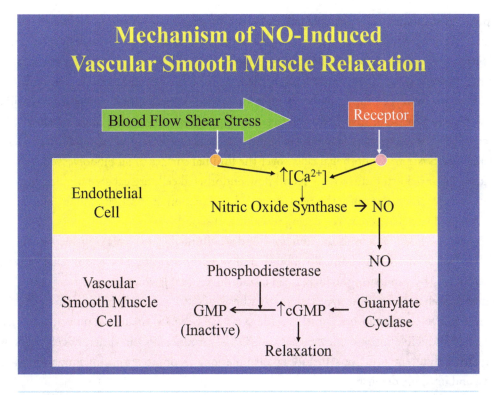

Fig. 6.20. Nitric oxide as an endothelium-derived relaxing factor.

despite an increase in arterial pressure within the autoregulatory range by proposing that an increase in arterial pressure will lead to oversupply of blood flow and oxygen for cellular metabolism, causing a decrease in the release of vasodilator by cells and arteriolar vasoconstriction until oxygen supply matches oxygen demand.

A major challenge to the vasodilator hypothesis is that the putative vasodilator that mediates autoregulation of blood flow has not been definitively identified. Myogenic hypothesis is an alternative hypothesis that explains autoregulation of blood flow by proposing that the increase in vascular resistance, in response to an increase in arterial pressure, is caused by mechanical stretch-induced activation of vascular smooth muscle cells.

Exercise hyperemia is a physiological response of skeletal muscle to exercise, in which muscle blood flow increases spontaneously during exercise to the level that is necessary for supplying oxygen to muscle metabolism, relatively independent of the autonomic nervous system. The vasodilator hypothesis explains exercise hyperemia by proposing that an increase in muscle metabolism during exercise leads to an increase in oxygen demand, which, if not met by an increase in oxygen supply, will cause an increase in the release of vasodilator by muscle cells. The accumulation of vasodilator will then lead to arteriolar vasodilation and an increase in blood flow until oxygen supply matches oxygen demand.

Hypoxic vasodilation is a physiological response of almost all organs to hypoxia (insufficient supply of oxygen for metabolism), in which organ blood flow increases spontaneously during hypoxia to the level that is necessary for supplying oxygen to organ metabolism, relatively independent of the autonomic nervous system. The vasodilator hypothesis explains hypoxic vasodilation by proposing that insufficiency in oxygen supply—for example, due to low arterial blood oxygenation—will cause an increase in the release of vasodilator by hypoxic cells. The accumulation of vasodilator will then lead to arteriolar vasodilation and increase in blood flow until oxygen supply matches oxygen demand.

Vasodilator System. Endothelial cells cover the luminal surface of all blood vessels. In addition to providing an anticlotting surface for blood flow, endothelial cells are capable of releasing relaxing and contracting factors that cause the relaxation and contraction of vascular smooth muscle cells in the wall of blood vessels, thereby regulating resistance of blood vessels.

Nitric oxide (NO) is a well-documented **endothelium-derived relaxing factor.** As shown in Fig. 6.20, shear stress (mechanical rubbing) and receptor activation of endothelial cells induce the release of NO by stimulating an increase in intracellular $[Ca^{2+}]$, which then activates the Ca^{2+}-dependent endothelial nitric oxide synthase for synthesis of NO from L-arginine. Three isoforms of nitric oxide synthases—endothelial nitric oxide synthase (eNOS), neuronal nitric oxide synthase (nNOS), and inducible nitric oxide synthase (iNOS)—produce NO in different cell types, where NO performs different functions including vasodilation, vasoprotection, inflammation, synaptic plasticity, and immune defense.

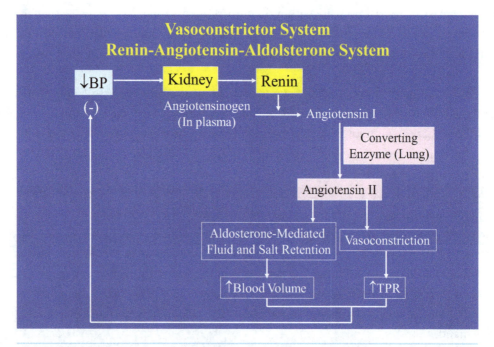

Fig. 6.21. Renin-angiotensin aldolsterone system as a vasoconstrictor system.

NO is a lipid-soluble gas that can diffuse across the cell membrane of vascular smooth muscle cells into the cytoplasm, where NO activates guanylyl cyclase, an enzyme that catalyzes the synthesis of cyclic GMP (cGMP) from GTP. Cyclic GMP is the intracellular molecule that induces vascular smooth muscle relaxation by activating cGMP-dependent kinase and other effector proteins. Termination of NO-induced vascular smooth muscle relaxation is mediated by the degradation of cyclic GMP to GMP by phosphodiesterase.

Penile erection is mediated by an increase in penile blood flow, mediated by NO-induced vascular smooth muscle relaxation. As pointed out in the last paragraph, enzymatic activities of NO-activated guanylyl cyclase and phosphodiesterase together determine the intracellular [cGMP] that mediates vascular smooth muscle contraction. Phosphodiesterase-5 has been identified as the phosphodiesterase isoform that catalyzes the breakdown of cGMP in penile arteries. For the treatment of erectile dysfunction, pharmacologic inhibitors of phosphodiesterase-5 have been developed for enhancing the level of cGMP in penile vascular smooth muscle cells during an erection.

Endothelial cells also release endothelium-derived contracting factors that induce contraction of vascular smooth muscle cells. Endothelin is a well-documented endothelium-derived contracting factor. Specific physiological functions of endothelium-derived contracting factors remain unclear.

Vasoconstrictor System. Renin-Angiotensin-Aldosterone System—consisting of an enzyme (renin), a peptide (angiotensin), and a steroid hormone—is an important regulator of arterial blood

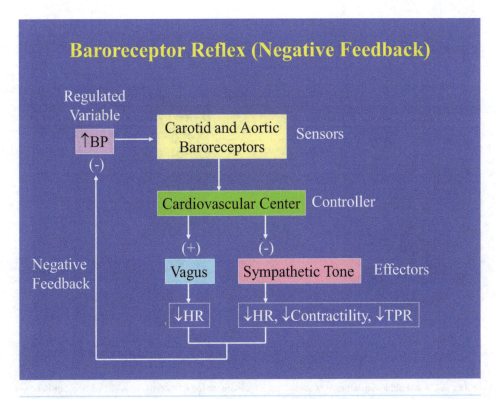

Fig. 6.22. Baroreceptor reflex.

pressure. The renin-angiotensin-aldosterone system controls blood pressure by regulating vascular resistance and blood volume in the systemic circulation. Fig. 6.21 shows the function of the renin-angiotensin-aldosterone system as a negative feedback mechanism for the regulation of blood pressure. As shown in Fig. 6.21, in response to a decrease in blood pressure and renal blood flow, specific cells in the kidneys—juxtaglomerular cells—secrete renin into the circulation. Renin is a proteolytic enzyme that catalyzes the breakdown of angiotensinogen—a protein in the plasma—to angiotensin I, a relatively inactive peptide. Angiotensin I is then transported to the pulmonary circulation, where an endothelial cell surface enzyme—angiotensin-converting enzyme—catalyzes the breakdown of angiotensin I to angiotensin II—a highly active peptide. Angiotensin II induces vasoconstriction in almost all organ systems by activating angiotensin II receptors on vascular smooth muscle, thereby increasing vascular resistance in the systemic circulation (total peripheral resistance). Angiotensin II stimulates the release of a steroid hormone—aldosterone—by adrenocortical cells in the adrenal gland. Aldosterone stimulates renal reabsorption of Na^+, thereby increasing the blood volume. By increasing total peripheral resistance and blood volume, the renin-angiotensin-aldosterone system restores blood pressure toward the normal level.

Pharmacologic inhibitors of the renin-angiotensin-aldosterone system have been developed for the treatment of systemic hypertension—for example, angiotensin-converting enzyme (ACE) inhibitors, angiotensin II receptor antagonists, and aldosterone receptor antagonists.

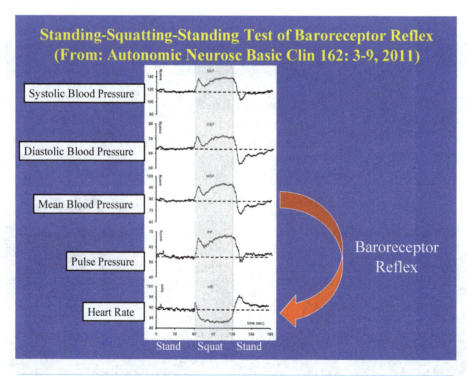

Fig. 6.23—Standing-Squatting-Standing Test of Baroreceptor Reflex. (From: Philips and Scheen. Autonomic Neurosc Basic Clin 162: 5, 2011).

Cardiovascular Reflex. The baroreceptor reflex functions as a negative feedback mechanism for short-term stabilization of blood pressure in response to external perturbations—for example, a sudden change in posture. As shown in Fig. 6.22, arterial blood pressure is sensed by baroreceptors—carotid baroreceptors situated at the bifurcation of common carotid arteries and aortic baroreceptors at the aortic arch—that transmit sensory signals to the cardiovascular center in the brain stem. A sudden increase in blood pressure above a set point will cause the cardiovascular center to increase parasympathetic (vagal) output and decrease sympathetic output to the heart and vascular system, resulting in decreases in heart rate, cardiac contractility, and total peripheral resistance, thereby causing blood pressure to decrease toward the set point. Conversely, a sudden decrease in blood pressure below a set point will cause the cardiovascular center to decrease parasympathetic (vagal) output and increase sympathetic output to the heart and vascular system, resulting in increases in heart rate, cardiac contractility, and total peripheral resistance, thereby causing blood pressure to increase toward the set point.

Fig. 6.23 demonstrates the response of the baroreceptor reflex to posture-induced changes in arterial blood pressure. As shown in Fig. 6.23, a sudden change in posture from standing to squatting causes rapid increases in systolic arterial pressure, diastolic arterial pressure, mean arterial pressure, and pulse pressure, accompanied by a concomitant decrease in heart rate. The increase in arterial blood pressure is caused by an increase in circulating blood volume as a result of squatting-induced compression of peripheral veins. The concomitant decrease in heart rate reflects the response from the baroreceptor reflex to the sudden increase in arterial pressure—decrease in sympathetic output and increase in vagal output from the cardiovascular center to the heart. By lowering heart rate, the baroreceptor reflex moderates the increase in arterial blood pressure caused by the sudden change in posture from standing to squatting.

Fig. 6.23 shows the effect of returning the posture from squatting to standing on the circulation. As shown in Fig. 6.23, changing the posture from squatting to standing causes rapid decreases in systolic arterial pressure, diastolic arterial pressure, mean arterial pressure, and pulse pressure, accompanied by an increase in heart rate. The decrease in blood pressure is caused by a decrease in circulating blood volume as a result of gravity-induced expansion of veins in the lower extremities. The concomitant increase in heart rate reflects the response of the baroreceptor reflex to the sudden decrease in arterial blood pressure—increase in sympathetic output and decrease in vagal output from the cardiovascular center to the heart. By increasing heart rate, the baroreceptor reflex moderates the decrease in arterial blood pressure caused by the sudden change in posture from squatting to standing. Arterial blood pressure and heart rate then gradually return to their steady-state values after redistribution of blood volume throughout the circulation.

As shown in Fig. 6.23, the baroreceptor reflex functions as a buffer of arterial pressure by regulating sympathetic and vagal output from the cardiovascular center to the heart and vascular system in response to changes in arterial pressure caused by external perturbations such as postural change. Experimental studies have shown that elimination of the baroreceptor reflex causes wide fluctuations in arterial pressure during a twenty-four-hour period. Integrity of the baroreceptor reflex in a patient can be tested clinically using the tilt-table test. The posture of a patient secured to a tilt

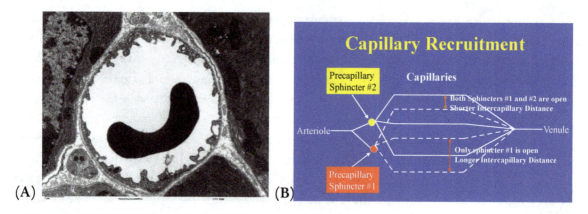

Fig. 6.24—A. Transmission electron microscopic image of a capillary containing a red blood cell.   **B.** Capillary recruitment for regulating intercapillary distance for diffusional transport.

table can be changed rapidly between supine (lying down) position and upright (standing) position. Patients having a dysfunctional baroreceptor reflex may faint when the posture is suddenly changed from the supine to an upright position.

The baroreceptor reflex adapts to long-term change in arterial pressure by shifting its set point in parallel with mean arterial pressure. This characteristic of the baroreceptor reflex has two physiological implications. First, the baroreceptor reflex is always active in moderating changes in arterial pressure in both normotensive and hypertensive individuals. Second, the baroreceptor reflex is generally not a cause of arterial hypertension.

Capillaries. Capillaries function as diffusional transport center. Capillaries—the smallest and thinnest blood vessels in the vascular system—consist of a single layer of endothelial cells. The primary function of capillaries is to provide a large surface area and short distance for diffusional exchange between blood and cells. As shown in Fig. 6.24A, the diameter of a capillary is similar to the diameter of a red blood cell, thereby minimizing the diffusion distance between the red blood cell inside the capillary and cells surrounding the capillary. In all organs, the capillary wall is highly permeable to lipid-soluble molecules—for example, oxygen and carbon dioxide. In most organs, with the exception of the central nervous system, gaps between endothelial cells allow the diffusion of charged and moderately large molecules—for example, ions and glucose—across the capillary wall. An important exception is the central nervous system, where capillaries are formed by tightly connected endothelial cells that constitute the blood-brain barrier to the diffusion of most molecules. Transport of most molecules across the blood-brain barrier is mediated by carrier-mediated transporters, endocytosis, and exocytosis.

The distance between neighboring capillaries—intercapillary distance—is an important determinant of diffusional transport between a capillary and the surrounding cells. Half of intercapillary distance is the maximum diffusion distance between a capillary and cells surrounding a capillary. Precapillary sphincters regulate intercapillary distance by controlling the number of capillaries that are open for perfusion—a process known as capillary recruitment. Precapillary sphincters are terminal arterioles containing a single layer of vascular smooth muscle cells. Relaxation of vascular smooth muscle cells in a precapillary sphincter leads to the opening of the sphincter and opening

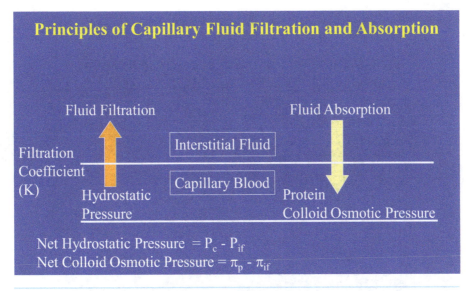

Fig. 6.25. Principles of capillary fluid filtration and absorption.

of capillaries for perfusion, whereas contraction of vascular smooth muscle cells in a precapillary sphincter leads to the closing of capillaries for perfusion. By reducing the diffusion distance, capillary recruitment is an important mechanism for enhancing diffusional transport. For example, Fig. 6.24B shows two precapillary sphincters that control the opening of two separate interdigitating networks of capillaries with similar intercapillary distances. As shown in Fig. 6.24B, simultaneous relaxation of the two precapillary sphincters together leads to opening of the two capillary networks and a 50 percent reduction of the intercapillary distance for diffusion.

Capillary Fluid Filtration and Absorption. In most organs other than the central nervous system, gaps between endothelial cells lining the capillary wall allow the exchange of water, ions, and small molecules—for example, Na^+, K^+, Cl^-, glucose—between the capillary plasma and the surrounding interstitial fluid. Gaps between endothelial cells are generally too small for the passage of proteins, resulting in the retention of protein in the plasma.

As shown in Fig. 6.25, the relatively high concentration of protein and therefore high colloid osmotic pressure in the plasma (π_c) and the relatively low protein concentration and therefore low colloid osmotic pressure in the surrounding interstitial fluid (π_{if}) together generate a colloid osmotic pressure gradient across the capillary wall, as shown in the following equation:

$$\text{Colloid Osmotic Pressure Gradient} = \pi_c - \pi_{if}$$

Counteracting the colloid osmotic pressure gradient is the hydrostatic pressure gradient that is generated by the relatively high blood pressure in the capillary (P_c) and the relatively low hydrostatic pressure in the surrounding interstitial fluid (P_{if}), as shown in the following equation:

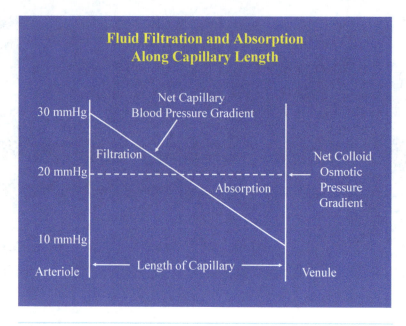

Fig. 6.26. Fluid filtration and absorption along capillary length.

$$\text{Hydrostatic Pressure Gradient} = P_c - P_{if}$$

The hydrostatic pressure gradient favors the diffusion of water from the capillary to the interstitium, whereas the colloid osmotic pressure gradient favors diffusion of water from the interstitium to the capillary. Net fluid movement across a capillary wall is determined by the difference between the two pressure gradients and capillary permeability (K), as shown in the following equation:

$$\text{Fluid Movement Across Capillary Wall} = (P_c - P_{if} - \pi_c + \pi_{if}) \times K$$

Fig. 6.26 shows the levels of net hydrostatic pressure gradient and net colloid osmotic gradients along the length of a capillary in most organs—for example, skeletal muscle and cardiac muscle—where capillary permeability is relatively small. The decrease in net hydrostatic pressure along the length of a capillary reflects the decrease in capillary blood pressure caused by the capillary resistance. The relatively constant net colloid osmotic pressure along the length of a capillary reflects the relatively small change in capillary volume due to the low capillary permeability. As shown in Fig. 6.26, fluid filtration occurs at the beginning portion of a capillary, where net hydrostatic pressure is higher than net colloid osmotic pressure. In comparison, fluid absorption occurs at the end portion of a capillary, where net hydrostatic pressure is lower than net colloid osmotic pressure. In most organs—for example, skeletal muscle and cardiac muscle—net fluid movement across the capillary wall is relatively small due to the relatively low capillary permeability. This is not the case in the glomerular filtration apparatus in the kidney, where the high capillary permeability of the glomerular capillary allows

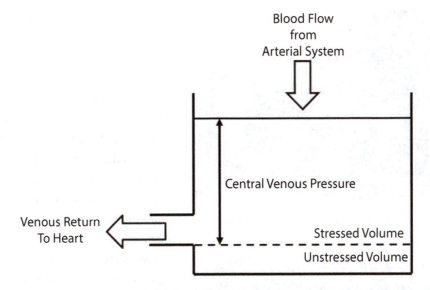

Fig. 6.27—Venous System Functions as a Volume Reservoir. Unstressed volume represents venous volume that does not contribute to central venous pressure. Stressed volume represents venous volume that contributes to central venous pressure for driving venous return to the heart.

filtration of fluid from the glomerulus into the renal tubular system. The mechanism of glomerular filtration will be covered in detail in the chapter on renal physiology.

Veins. The venous system functions as a volume reservoir in the cardiovascular system, because the high compliance of veins allows the accumulation of large blood volume in the venous system without a large increase in venous blood pressure. Approximately 70 percent of the total blood volume is stored in the venous system, but the venous pressure is normally very low. For example, central systemic venous pressure is near zero, and central pulmonary venous pressure is 2–3 mmHg. Fig. 6.27 illustrates the function of the venous system as a volume reservoir, using a water storage tank as a model. The unstressed venous volume below the water outlet represents the blood volume that can be held by the venous system before full expansion of the venous system that is necessary for any increase in venous pressure. The stressed venous volume represents the venous volume above the unstressed venous volume. The large diameter of the storage tank represents the large compliance of the venous system that enables the storage of a large venous volume with a small increase in venous pressure. Central venous pressure (CVP) for driving venous return to the heart is determined by the stressed venous volume and the venous compliance.

During exercise, the venous system functions as a volume reservoir for increasing circulating blood volume, when sympathetic stimulation-induced venous constriction decreases unstressed venous volume and venous compliance, thereby shifting blood from the venous system to the circulation for increasing venous return, ventricular filling, and cardiac output. In severe hemorrhage, venous constriction also plays a significant role in restoring cardiac output by shifting venous blood into the circulation.

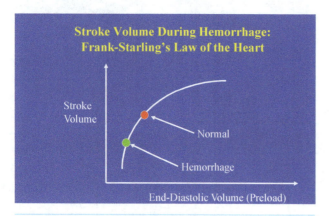

Fig. 6.28—Hemorrhage-induced decrease in stroke volume due to a decrease in end-diastolic volume.

(A)

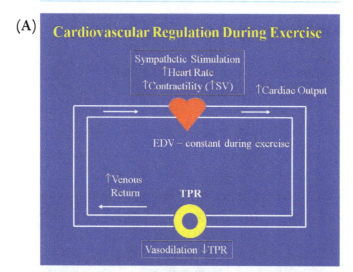

(B)

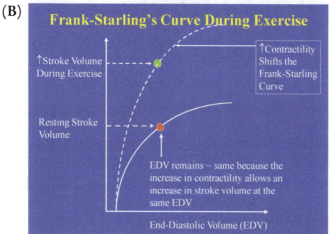

Fig. 6.29—**A.** Vasodilation of arterioles in exercising muscle lowers total peripheral resistance to increase muscle blood flow, which must be provided by an increase in cardiac output. **B.** Sympathetic stimulation of the heart during exercise increases cardiac contractility and thereby shifts the Frank-Starling curve to the left and upward direction.

Orthostatic hypotension is the drop in blood pressure that occurs in the standing posture. In orthostatic hypotension, the venous system behaves as a volume sink for decreasing circulating blood volume, when gravity causes a shifting of blood from the circulation to veins in the lower extremities of a person in the standing posture, thereby causing decreases in circulating volume, cardiac output, and blood pressure. Fainting caused by orthostatic hypotension is a physiological response, because fainting often causes a person to lie down, thereby facilitating return of venous blood from the lower extremities to the circulation for the restoration of cardiac output and blood pressure. A fainting person should be allowed to lie down until the person regains consciousness and receives medical assistance.

INTEGRATIVE CARDIOVASCULAR PHYSIOLOGY AND PATHOPHYSIOLOGY

The following analysis illustrates how the heart and the vascular system function together to support cardiac output and arterial blood pressure in response to three challenges to the cardiovascular system—severe loss of blood volume in severe hemorrhage, significant increase in oxygen demand in intense endurance exercise, and significant decrease in cardiac contractility in congestive heart failure.

Hemorrhage. The significant loss of blood volume in severe hemorrhage can lead to significant decreases in ventricular end-diastolic volume, stroke volume, and cardiac output, as predicted by the Frank-Starling law of the heart (Fig. 6.28). In response to severe hemorrhage, the cardiovascular system attempts to maintain arterial blood pressure for the perfusion of critical organs—heart and brain—for immediate survival by increasing heart rate, cardiac contractility, total peripheral resistance, and venous return. The increase in heart rate is mediated by sympathetic stimulation of the sinoatrial node and conductive system, and the increase in cardiac contractility is mediated by sympathetic stimulation of cardiac muscle cells. The increase in total peripheral resistance is mediated by preferential vasoconstriction of arterioles in organs that are most sensitive to sympathetic stimulation—kidneys, gastrointestinal tract, and skin. The increase in venous return is caused by sympathetic stimulation of the venous system, thereby shifting some blood volume from the venous system to the circulation for enhancing venous return to the heart, thereby enhancing ventricular filling. In addition, the decrease in capillary blood pressure, as a result of a decrease in arterial blood pressure, leads to an increase in fluid absorption from the interstitial fluid to the capillary and moderately increasing the circulating blood volume. In a hospital setting, hemorrhage can be treated by transfusion. In disaster situations when blood is not available, hemorrhage may be managed by intravenous infusion of isosmotic fluid—for example, normal saline (0.9% NaCl).

Intense Endurance Exercise. The high mechanical power output from skeletal muscle cells during intense endurance exercise is driven by aerobic metabolism that is dependent on the supply of oxygen by cardiac output. As shown in Fig. 6.29A, intense vasodilation in exercising muscles causes a significant decrease in total peripheral resistance in the systemic circulation and a significant

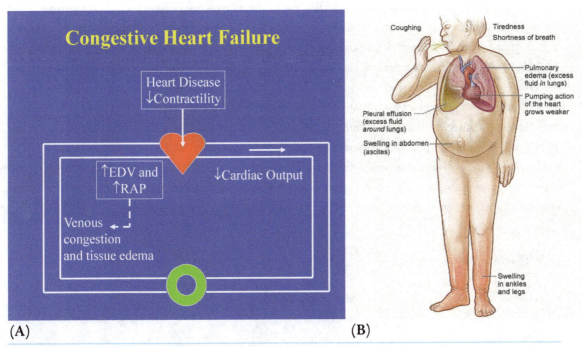

(A) (B)

Fig. 6.30. Congestive heart failure.

increase in venous return to the heart. To maintain normal arterial pressure during intense exercise, the heart increases cardiac output to match the increase in venous return by increasing heart rate and stroke volume. Heart rate can increase by up to 300 percent, as a result of sympathetic stimulation of the sinoatrial node and conductive system. In comparison, stroke volume can increase by up to 50 percent, as a result of an increase in cardiac contractility, mediated by sympathetic stimulation of cardiac muscle cells. As shown in Fig. 6.29B, sympathetic stimulation of cardiac muscle cells increases cardiac contractility, as indicated by a shifting of the Frank-Starling curve to the left and upward direction, resulting in an increase in stroke volume at a given end-diastolic volume. In addition, sympathetic stimulation of the venous system and compression of the veins by muscle pumping together shift some blood from the venous system to the circulation to increase ventricular filling and end-diastolic volume for stroke volume production. To direct cardiac output preferentially to the organs essential for exercise performance, the cardiovascular system significantly decreases blood flow to the kidney and gastrointestinal tract during intense exercise by sympathetic stimulation of arteriolar vasoconstriction in these two organs.

Congestive Heart Failure is characterized by a significant decrease in cardiac output, accompanied by increase in systemic venous pressure. As shown in Fig. 6.30A, the primary cause of congestive heart failure is the decrease in cardiac contractility, which lowers the ability of the heart to pump stroke volume into the circulation. Patients having cardiac failure are lethargic, because the low cardiac output limits the patient's ability to pursue normal physical activities. A typical compensatory response of the cardiovascular system is to direct the limited cardiac output to critical organs for immediate survival—for example, the heart and brain—and reduce blood flow to other organs—for example, the kidneys and gastrointestinal tract. The kidneys respond to a decrease in renal blood flow by increasing the reabsorption of Na^+ and water, thereby causing expansion of extracellular fluid volume and an increase in central venous pressure. This compensatory response from the kidneys is ill suited for survival for several reasons. First, an enlarged heart filled with an excessively large end-diastolic volume is energetically inefficient for generating cardiac output, causing an increase in cardiac oxygen consumption and further deterioration of the failing heart. Second, an increase in central venous pressure as a result of extracellular fluid volume causes an increase in capillary hydrostatic pressure, which drives fluid movement from capillaries into interstitial space. As shown in Fig. 6.30B, fluid accumulation causes swelling of the abdomen and lower extremities. Fluid accumulation in the alveolar space in the lungs causes pulmonary edema, shortness of breath, and coughing. Cardiac failure patients are often treated with diuretics to decrease fluid congestion in the extracellular space, and positive inotropic agents to enhance cardiac contractility.

KEY TERMS

- active filling
- afterload
- aortic valve

- arteries
- arteries
- arterioles

- arterioles
- atrioventricular node
- atrium

- augmented ECG leads
- autoregulation
- baroreceptor reflex
- blood pressure
- bundle of His
- capillaries
- cardiac action potential
- cardiac cycle
- cardiac muscle
- cardiac output
- colloid osmotic pressure
- contractility
- diastolic pressure
- ejection fraction
- electrical axis of the heart
- electrocardiogram
- end-diastolic volume
- end-systolic volume

- endothelium-derived relaxing factor
- endurance exercise
- exercise hyperemia
- HCN channels
- heart rate
- hemorrhage
- hydrostatic pressure
- hypoxic vasodilation
- isovolumic contraction
- isovolumic relaxation
- mean arterial pressure
- mitral valve
- nitric oxide
- pacemaker potential
- passive filling
- phonocardiogram
- Poiseuille's law

- precordial ECG leads
- preload
- pulmonary circulation
- pulmonary valve
- pulse pressure
- purkinje fibers
- renin-angiotensin-aldosterone system
- sinoatrial node
- standard ECG leads
- stroke volume
- systemic circulation
- systolic pressure
- tricuspid valve
- vascular resistance
- veins
- ventricle

IMAGE CREDITS

7

RESPIRATORY PHYSIOLOGY

The continued supply of oxygen to and removal of carbon dioxide from the body by the respiratory system is essential for survival. Respiratory physiology is concerned with the mechanisms for transporting two molecules—oxygen and carbon dioxide—between the atmospheric air and cells in the body. Lungs transport oxygen from air into blood, and the circulation carries oxygen from the lungs to cells, as shown in the following scheme:

Oxygen: Air → Lungs → Circulation → Cells

In the reverse direction, the circulation carries carbon dioxide from cells to the lungs, and the lungs transport carbon dioxide to the air, as shown in the following scheme:

Carbon Dioxide: Air ← Lungs ← Circulation ← Cells

The two major topics of respiratory physiology are **mechanics of lung ventilation** for moving air into and out of lungs, and **blood transport of oxygen and carbon dioxide.**

LEARNING OBJECTIVES

1. **Lung Volumes and Capacities.** Describe lung volumes and capacities that are typically measured during spirometry, and explain their physiological significance.

2. **Breathing Mechanism.** Explain the regulation of lung volume by intrapleural pressure during inspiration and expiration; define lung compliance, and discuss the contributions of tissue elasticity and surface tension to lung compliance; define airway resistance and discuss its physiological significance; compare and contrast obstructive and restrictive lung diseases in terms of cause and diagnosis.

3. **Alveolar Ventilation.** Define alveolar ventilation, and compare and contrast deep breathing and shallow breathing in terms of their effects on alveolar ventilation at the same minute ventilation.

4. **Regulation of Alveolar PO_2 and PCO_2.** Describe the quantitative dependencies of alveolar PO_2 and alveolar PCO_2 on inspired air composition, alveolar ventilation, and metabolic rate; describe the quantitative relationship between alveolar PO_2 and PCO_2.

5. **Blood Gas Transport.** Define the Fick principle, and discuss its utility in explaining how cardiac output and arterial–venous difference in oxygen content together support oxygen consumption rate; discuss the biochemical basis of the oxyhemoglobin dissociation curve and its regulation by chemical modulators; describe the three mechanisms of blood CO_2 transport; explain the specific functions of hemoglobin in blood CO_2 transport.

6. **Regulation of Ventilation.** Describe the central and peripheral chemoreceptor mechanisms by which plasma PO_2, PCO_2, and $[H^+]$ regulate ventilation.

7. **Adaptation to High Altitude.** Describe the mechanism by which high altitude challenges survival and the compensatory mechanisms by which the respiratory system adapts to high altitude.

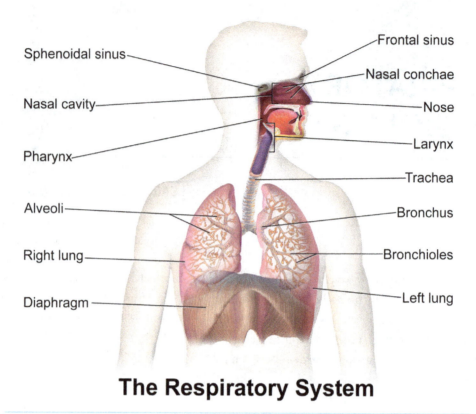

Sphenoidal sinus
Nasal cavity
Pharynx
Alveoli
Right lung
Diaphragm

Frontal sinus
Nasal conchae
Nose
Larynx
Trachea
Bronchus
Bronchioles
Left lung

The Respiratory System

Fig. 7.1. Respiratory system consisting of conducting airways and alveoli at the end for gas exchange with blood in the surrounding capillary.

OVERVIEW OF THE RESPIRATORY SYSTEM

As shown in Fig. 7.1, the airway system resembles a tree in its branching from trachea to right and left bronchi, small bronchioles, and eventually alveoli—microscopic air sacs made up of a monolayer of airway epithelial cells. As shown in Fig. 7.2, pulmonary arteries and veins follow the airway system to form a capillary network around the alveoli, where oxygen and carbon dioxide are exchanged rapidly between alveolar air and pulmonary capillary blood by diffusion.

LUNG VOLUMES AND CAPACITIES

Measurements of lung volumes are important for differentiating types of lung diseases; for example, the collapse of a lung in pneumothorax can lead to a substantial decrease in total lung capacity. Lung volumes can be measured by a spirometer, a device that measures the volumes of air moving into and out the lungs during inspiration and expiration, as shown in Fig. 7.3A. During spirometry, a person's mouth is connected to a spirometer by a tube and the subject's nose is clipped to prevent the movement of air through the nostrils.

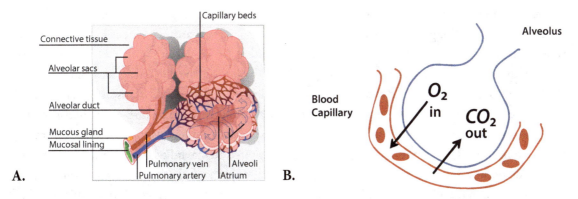

A.

B.

Fig. 7.2—A. Structure of alveolus and the surrounding pulmonary capillary network.   **B.** Gas exchange between alveolar air and capillary blood.

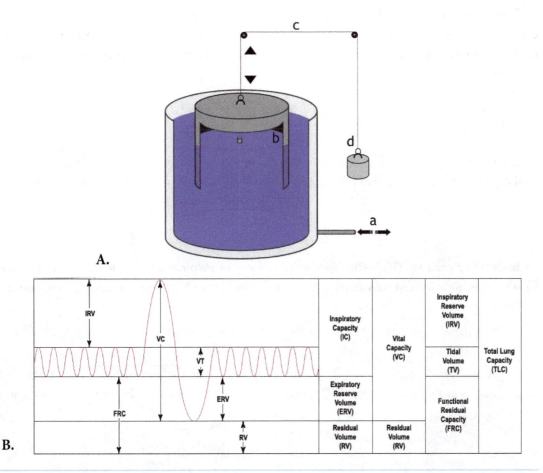

A.

B.

Fig. 7.3—A. Spirometer for measuring lung volumes. A flexible air tube connects a subject's mouth to the port of the spirometer (a), which is connected to the inside of a drum (b). The drum is connected to a balance weight (d) by wire and pulley (c). Breathing by the subject results in the movement of air in and out of the drum, causing the up and down movement of the drum, which is recorded as lung volumes on a chart, as shown in panel B. The spirometer shown in panel A is useful for understanding lung volumes. In clinics, lung volumes are measured using electronic spirometers without the drum and pulley system.   **B.** Lung volumes recorded by the spirometer.

Lung Volumes. Fig. 7.3B shows a spirometric recording of changes in lung volume with time during several breathing cycles. In a spirometric recording, an upward movement indicates inspiration, whereas a downward movement indicates expiration. As shown in Fig. 7.3B, during the first series of quiet breathing cycles, tidal volume (TV) is the volume of air that enters into and exits from the lungs during inspiration and expiration. As shown in Fig. 7.3B, a maximal inspiration reveals the size of the inspiratory reserve volume (IRV)—the lung volume that can be inspired to maximize the tidal volume. Similarly, as shown in Fig. 7.3B, a maximal expiration reveals the size of the expiratory reserve volume (ERV)—the lung volume that can be expired to maximize the volume of expiration.

As shown in Fig. 7.3B, residual volume (RV) is the volume of air that remains in the lungs at the end of maximal expiration. Residual volume cannot be measured by spirometry, because residual volume cannot be expired from the lungs. Residual volume can be measured by dilution with a tracer gas.

Lung Capacities. Each lung capacity is a sum of two or more lung volumes. For example, vital capacity (VC)—maximum tidal volume of ventilation—is a sum of inspiratory reserve volume (IRV), normal tidal volume (TV), and expiratory reserve volume (ERV), as shown in the following equation:

$$VC = IRV + TV + ERV$$

Total lung capacity (TLC)—the total amount of air in a maximally inflated lung—is a sum of vital capacity (VC) and residual volume (RV), as shown in Fig. 7.3B and the following equation:

$$TLC = VC + RV$$

Inspiratory capacity (IC)—the maximum volume of inspiration—is the sum of tidal volume (TV) and inspiratory reserve volume (IRV), as shown in Fig. 7.3B and the following equation:

$$IC = TV + IRV$$

Functional residual capacity (FRC)—the lung volume at the end of quiet breathing—is the sum of expiratory reserve volume (ERV) and residual volume (RV), as shown in Fig. 7.3B and the following equation:

$$FRC = ERV + RV$$

Normal ranges of lung volumes and lung capacities for an individual can be estimated from height, body weight, and sex. Lung diseases can lead to abnormalities in lung volumes. For example, the collapse of a lung in pneumothorax would result in a substantial decrease in vital capacity. Lung hyperinflation caused by airway obstruction in asthmatic patients can lead to a significant decrease in inspiratory reserve volume and a significant increase in expiratory reserve volume.

BREATHING MECHANISM

A breathing cycle consists of the movement of air into and out of the lungs, which are driven by the difference between **atmospheric pressure** and **alveolar pressure**, as shown in the following equation:

$$\text{Airflow} = (\text{Atmospheric Pressure} - \text{Alveolar Pressure}) / \text{Airway Resistance}$$

As shown in the above equation, alveolar pressure is the variable that drives airflow into and out of the lungs. Alveolar pressure is dependent on the balance between two forces—**lung recoil** driving lung collapse, and **transpulmonary pressure** driving lung expansion. Lung recoil consists of the elastic force of connective tissues and surface tension at the air-liquid interface on the alveolar surface. Lung recoil increases with lung volume, as shown in the following relation:

$$\uparrow \text{Lung Volume} \rightarrow \uparrow \text{Lung Recoil}$$

Transpulmonary pressure is the pressure difference across the wall of a lung, as shown in the following equation:

$$\text{Transpulmonary Pressure} = \text{Alveolar Pressure} - \text{Intrapleural Pressure}$$

Intrapleural Pressure is the pressure in the pleural space between the lung surface and chest wall, as shown in Fig. 7.4B. In a healthy person, the pleural space is filled with a small amount of fluid without any air (Fig. 7.4B). The pleural space becomes evident in a patient having pneumothorax, when the pleural space becomes leaky and filled with air.

Alveolar Pressure is Zero at Stable Lung Volumes. At stable lung volumes—reached at the end of inspiration and expiration—lung recoil is balanced by transpulmonary pressure, alveolar pressure is zero, and airflow is zero. For example, at the beginning of inhalation (Fig. 7.4A, **time zero**), alveolar pressure is zero, because the transpulmonary pressure ($P_{alv} - P_{pl} = 5$ cm H_2O) is balanced by lung recoil at the stable end-expiratory volume (2.5 liters). Similarly, at the end of inhalation (Fig. 7.4A, **mid-time**), alveolar pressure is zero, because the transpulmonary pressure (8 cm H_2O) is balanced by lung recoil at the stable end-inspiratory volume (3 liters).

Alveolar Pressure is Negative During Inspiration. As shown in Fig. 7.4A, the beginning of inspiration—increase in lung volume—is driven by a decrease in alveolar pressure (P_{alv}) as a result of a decrease in intrapleural pressure (P_{pl}).

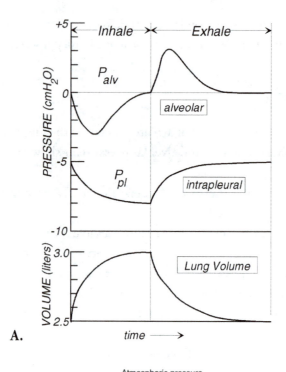

A.

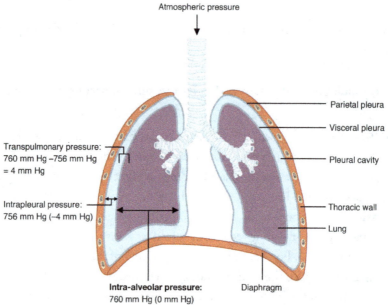

B.

Fig. 7.4—A. Changes in alverolar pressure, intrapleural pressure and lung volume during a breathing cycle.   **B.** Structure of pleura and pleural space.

$$\downarrow \text{Intrapleural Pressure} \rightarrow \downarrow \text{Alveolar Pressure} \rightarrow \text{Inspiration} (\uparrow \text{Lung Volume})$$

The decrease in intrapleural pressure during inspiration is caused by contractions of diaphragm and intercostal muscles, resulting in expansion of the thoracic cavity. The decrease in intrapleural pressure leads to an increase in transpulmonary pressure above lung recoil, resulting in negative alveolar pressure relative to atmospheric pressure and inspiration. The increase in lung volume during inspiration leads to an increase in lung recoil, until lung recoil reaches a level that balances the transpulmonary pressure, at which alveolar pressure becomes zero and lung volume stabilizes at the end-inspiratory volume (3 liters), as shown in Fig. 7.4A (mid-time).

Alveolar Pressure is Positive During Expiration. Mechanical events during expiration are the reverse of events during inspiration. As shown in Fig. 7.4A (mid-time), the beginning of expiration—decrease in lung volume—is driven by an increase in alveolar pressure that results from an increase in intrapleural pressure.

$$\uparrow \text{Intrapleural Pressure} \rightarrow \uparrow \text{Alveolar Pressure} \rightarrow \text{Expiration} (\downarrow \text{Lung Volume})$$

The increase in intrapleural pressure during expiration is caused by relaxation of diaphragm and intercostal muscles, resulting in a shrinking of the thoracic cavity. The increase in intrapleural pressure leads to a decrease in transpulmonary pressure to below lung recoil, resulting in positive alveolar pressure relative to atmospheric pressure and expiration. The decrease in lung volume during expiration leads to a decrease in lung recoil, until lung recoil reaches a level that balances transpulmonary pressure, at which alveolar pressure becomes zero and lung volume stabilizes at the end-expiratory volume (2.5 liters), as shown in Fig. 7.4A (end-time).

Lung Compliance is a measure of lung flexibility. Quantitatively, lung compliance is defined as the slope at any one point of the lung volume (V)-transpulmonary pressure (P) relation, as described by the following equation:

$$\text{Lung Compliance} = \Delta V / \Delta P$$

The above equation may be used to determine the change in transpulmonary pressure (ΔP) that is necessary for causing a change in lung volume (ΔV). As shown in Fig. 7.5A, lung compliance is highest at intermediate lung volume, which explains why least effort is needed for breathing at intermediate lung volumes. In comparison, lung compliance is low at very small and very large lung volumes. Lung compliance is inversely related to mechanical forces that tend to collapse the lungs. For a spherical structure, Laplace's law determines the transmural pressure (ΔP) necessary for stabilizing the structure at a radius (R) against a wall tension (T) that tends to collapse the structure.

$$\text{Laplace's Law:} \quad \Delta P = 2T/R$$

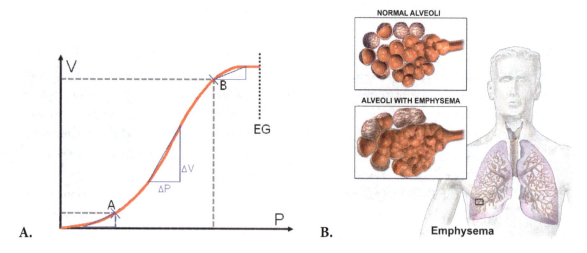

Fig. 7.5—A. Compliance is the slope (ΔV/ΔP) at one point of the volume–pressure relationship. Patients having emphysema exhibit abnormally high lung compliance due to the destruction of connective tissues in the lungs.

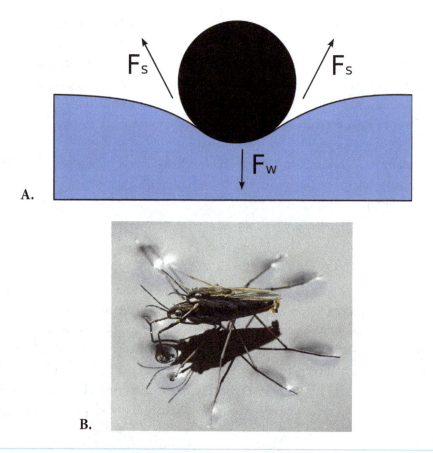

Fig. 7.6—A. Surface tension (Fs) resulting from adhesion between liquid molecules supports the weight of an object (Fw) on the liquid.   **B.** Surface tension supports the weight of insects on water.

TISSUE ELASTICITY AND SURFACE TENSION

Wall tension that tends to collapse the lungs consists of tissue elasticity and surface tension. Tissue elasticity consists of elasticity of protein filaments inside cells and connective tissues in the extracellular matrix. Excessive production of extracellular matrix in lung fibrosis can cause abnormally low lung compliance. Conversely, degradation of elastic fibers in smoking-induced lung emphysema can cause abnormally high lung compliance, as shown in Fig. 7.5B. Surface tension is the force at an air-liquid interface produced by the adhesion between liquid molecules. As shown in Fig. 7.6A, surface tension (Fs) can be sufficiently large to support the weight of an object (Fw) on the liquid surface.

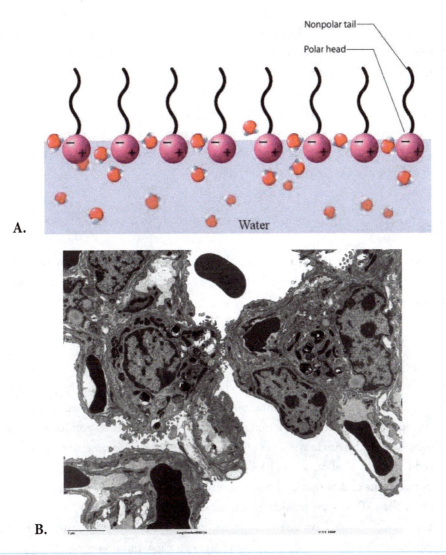

Fig. 7.7—A. Surfactant molecules—containing polar and non-polar ends—coats the water surface, thereby reducing the surface tension at the air-liquid interface. **B.** Alveolar surfactant is produced by type 2 alveolar cells—cells containing secretory granules.

Indentation of the liquid surface by an object results in a net force derived from surface tension that opposes the weight of the object. In nature, surface tension at the air-water interface is sufficiently large to support the relatively light weight of insects, as shown in Fig. 7.6B.

Water—the major component of body fluids—has a relatively high surface tension due to the high adhesive force between water molecules. As shown in Fig. 7.7A, water surface can be covered by amphiphilic molecules—with polar heads facing water and nonpolar tails facing air. Covering water surface by amphiphilic molecules can result in a reduction of surface tension due to the relatively weak adhesive force between nonpolar tails of amphiphilic molecules. In human lungs, pulmonary surfactant—consisting of amphiphilc molecules such as phospholipid and other molecules—covers the aqueous layer on the alveolar surface, thereby reducing the surface tension on the alveolar surface. Pulmonary surfactant is secreted by type 2 alveolar cells—tall and granule-containing cells of the alveolar wall, as shown in Fig. 7.7B. The other alveolar cell type—type 1 alveolar cell—is relatively flat in cell thickness, does not secrete pulmonary surfactant, and mostly provides mechanical integrity of the alveolar wall. Pulmonary surfactant is physiologically significant for enhancing lung compliance by reducing surface tension on the alveolar surface. Deficiency in pulmonary surfactant secretion in premature newborns can lead to difficulty in breathing, a condition known as infant respiratory distress syndrome. Synthetic and cow-derived pulmonary surfactants are available for the treatment of infant respiratory distress syndrome. In addition to lowering surface tension, pulmonary surfactant enhances host defense against air-borne pathogens.

AIRWAY RESISTANCE

As cited previously, air flow into and out of the lungs is driven by the difference between alveolar pressure and atmospheric pressure (ΔP) against airway resistance (R_{aw}), as described by the following equation:

$$Airflow = \Delta P / R_{aw}$$

Obstructive lung diseases such as asthma are characterized by abnormally high airway resistances and low airflow. A standard pulmonary function test for differentiating obstructive from restrictive diseases is to record lung volume as a function of time during forced expiration from total lung capacity. The data are used to calculate the FEV_1/FVC ratio—expired volume during the first one sec (FEV_1) divided by forced vital capacity (FVC). In this test, forced vital capacity is the difference in lung volume between the end of maximal inspiration and the end of forced expiration. A FEV_1/FVC ratio that is lower than 70 percent would suggest airway obstruction.

Obstructive and restrictive lung diseases can be differentiated clearly by measuring airflow and lung volume during cycles of maximal inspiration and expiration and presenting the data in the form of **flow volume loop**, as shown in Fig. 7.8. The abscissa (X-axis) in Fig. 7.8 represents expired air volume, ranging from zero, when the lung is fully inflated, to force vital capacity (FVC), when the

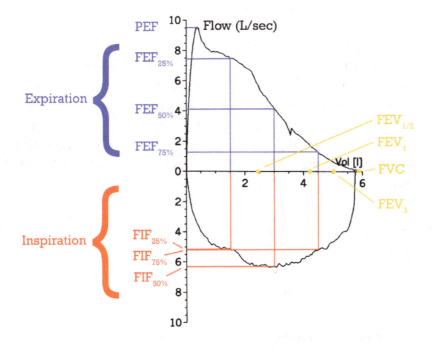

Fig. 7.8. Flow-volume loop during a respiratory cycle.

lung is fully deflated. The ordinate (Y-axis) in Fig. 7.8 represents airflow, ranging from negative values for inspiration to positive values for expiration. Accordingly, the bottom portion (negative airflow) of the loop represents inspiration from FVC toward zero. The top portion (positive airflow) of the loop represents expiration from 0 to FVC. As shown in Fig. 7.8, peak expiratory flow (PEF) is reached near the beginning of expiration. Obstructive lung disease is characterized by a significant decrease in peak expiratory flow. In comparison, restrictive lung disease is characterized by a significant decrease in lung volume with relatively small change in peak expiratory flow.

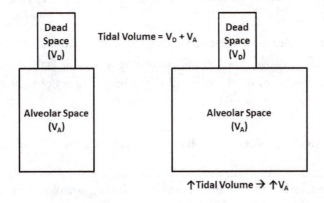

Fig. 7.9—Tidal volume consists of dead space (V_D) and alveolar space (V_A). Right panel shows the effect of increasing tidal volume on increasing alveolar space for ventilation.

DEAD SPACE AND RESPIRATORY UNITS

The airway system is necessary for carrying air between the atmosphere and gas exchange units (respiratory bronchioles and alveoli), whereas gas exchange occurs at the gas exchange units. The airway system that is not capable of gas exchange is considered anatomical dead space. In addition, gas exchange units that are not perfused by circulation are considered physiological dead space. Dead space does not contribute to gas exchange, because it retains alveolar air at the end of expiration, which is then mixed with inspired air for transport to the alveoli. Accordingly, tidal volume (TV) of each breath consists of dead space volume (V_D) and alveolar space volume (V_A), as illustrated in Fig. 7.9 and shown in the following equation:

$$TV = V_D + V_A$$

Minute Ventilation. Overall lung ventilation (minute ventilation) is the product of tidal volume and breathing frequency (f), as shown in the following equation:

$$\text{Minute Ventilation} = TV \times f$$

Values of tidal volume and breathing frequency vary with age, body size, gender, and metabolic activity. For an adult, typical values for tidal volume (TV) and breathing frequency (f) are 500 ml and 12/min, respectively. Accordingly, minute ventilation for this person can be calculated as follows:

$$\text{Minute Ventilation} = 500 \text{ ml} \times 12/\text{min} = 6,000 \text{ ml}/\text{min}$$

Alveolar Ventilation. To calculate ventilation of the gas exchange units (alveolar ventilation), it is necessary to subtract dead space ventilation from the overall lung ventilation (minute ventilation), as shown in the following equation:

$$\text{Alveolar Ventilation} = TV \times f - V_D \times f = (TV - V_D) \times f$$

Value of dead space varies with age, body size, gender, and cardiovascular health. For an adult, a typical value for dead space (V_D) is 150 ml. Accordingly, alveolar ventilation for this person can be calculated using this value of dead space and the previous values of tidal volume (500 ml) and respiratory frequency (12/min).

$$\text{Alveolar Ventilation} = (500 \text{ ml} - 150 \text{ ml}) \times 12/\text{min} = 4,200/\text{min}$$

Alveolar ventilation, rather than minute ventilation, is the physiologically significant ventilation for gas exchange. By choosing different combinations of tidal volume and breathing frequency, it is possible to change alveolar ventilation without changing minute ventilation.

Deep Breathing. As shown in the following calculation, by increasing tidal volume to 600 ml and decreasing breathing frequency to 10/min, it is possible to increase alveolar ventilation at the same minute ventilation (6000 ml/min):

$$\text{Alveolar Ventilation} = (600 \text{ ml} - 150 \text{ ml}) \times 10/\text{min} = 4{,}500 \text{ ml/min}$$

Shallow Breathing. As shown in the following calculation, by decreasing tidal volume to 400 ml and increasing respiratory frequency to 15/min, it is possible to decrease alveolar ventilation at the same minute ventilation (6,000 ml/min):

$$\text{Alveolar Ventilation} = (400 \text{ ml} - 150\text{ml}) \times 15/\text{min} = 3{,}750 \text{ ml/min}$$

In general, deep breathing improves alveolar ventilation. In addition, periodic deep inspirations are known to lower airway resistance by inducing airway dilation, but the underlying mechanism is not fully understood. Shallow breathing in the form of panting is used by some animals for heat dissipation.

REGULATION OF ALVEOLAR PCO$_2$ AND ALVEOLAR PO$_2$

Alveolar ventilation maintains physiological levels of blood PO_2 and PCO_2 by carrying atmospheric air—high in PO_2 and low in PCO_2—into alveoli and exhaling alveolar air—low in PO_2 and high in PCO_2—to the atmosphere. At sea level and moderate altitude, PO_2 and PCO_2 in each alveolus and its surrounding capillary blood are essentially identical, due to the rapid diffusion between alveolar air and capillary blood. Alveolar PO_2 and PCO_2 are determined by three variables: PO_2 and PCO_2 in inspired air, metabolic rates (O_2 consumption and CO_2 production), and alveolar ventilation.

Determinants of Alveolar PCO$_2$. Inspired PCO_2 is typically near zero, because atmospheric air contains a negligible amount of CO_2. Alveolar PCO_2 is then determined by only two variables: rate of metabolic CO_2 production and CO_2 removal by alveolar ventilation, as illustrated in the following scheme:

$$\text{Metabolic } CO_2 \text{ Production} \rightarrow \text{Alveolar } PCO_2 \rightarrow CO_2 \text{ Removal by Alveolar Ventilation}$$

The rate of CO_2 removal by alveolar ventilation equals the product of alveolar ventilation (V_A) and fraction of CO_2 in alveolar air (F_ACO_2), which is proportional to alveolar PCO_2, as shown in the following equation:

$$\text{Rate of } CO_2 \text{ Removal} = V_A \times F_ACO_2 = V_A \times P_ACO_2 / K$$

In the above equation, K is the proportional constant for PCO_2/FCO_2 ratio in alveolar air. In a steady state, the rate of metabolic CO_2 production (VCO_2) equals the rate of CO_2 removal.

$$VCO_2 = V_A \times P_ACO_2 / K$$
$$\text{Alveolar } PCO_2 \text{ Equation: } P_ACO_2 = K \times VCO_2 / V_A$$

The alveolar PCO_2 equation predicts that P_ACO_2 is determined by the VCO_2/V_A ratio. The alveolar PCO_2 equation implies that maintenance of normal level of P_ACO_2 depends on proportional matching of the metabolic rate by alveolar ventilation. For example, during exercise, the increase in CO_2 production rate must be matched by a proportional increase in alveolar ventilation. At sea level, P_ACO_2 is normally maintained at 40 mmHg by the respiratory system, because P_ACO_2 is a major determinant of blood pH. The determination of blood pH by P_ACO_2 and plasma bicarbonate concentration will be covered in detail later.

Systemic arterial blood PCO_2 is essentially identical to P_ACO_2, due to the rapid diffusion between air in an alveolus and blood in the surrounding capillary, as cited previously. Venous blood PCO_2 is organ-specific, because the ratio of organ metabolic rate/organ blood flow varies widely among body organs. For example, coronary venous PCO_2 is relatively high, because the metabolic rate of the heart is high relative to coronary blood flow. In comparison, renal venous PCO_2 is relatively low, because the metabolic rate of the kidneys is low relative to the high renal blood flow for clearing waste products. PCO_2 in mixed systemic venous blood from all organ systems is typically 45 mmHg.

Determinants of Alveolar PO_2. Alveolar PO_2 is determined by three variables: inspired PO_2, metabolic O_2 consumption rate, and alveolar ventilation, as illustrated in the following scheme:

$$O_2 \text{ Supply by Alveolar Ventilation} \rightarrow \text{Alveolar } PO_2 \rightarrow \text{Metabolic } O_2 \text{ Consumption}$$

Inspired PO_2 is a critical determinant of O_2 supply by alveolar ventilation, because inspired PO_2 reflects the oxygen content of inspired air and represents the upper limit of alveolar PO_2. During normal breathing of atmospheric air containing 21% O_2, inspired PO_2 (P_IO_2) entering alveoli can be calculated from barometric pressure (P_B) and the saturated water vapor pressure at 37°C (47 mmHg), as shown in the following equation:

$$P_IO_2 = 0.21 \times (P_B - 47)$$

At sea level, P_B is approximately 760 mmHg, and P_IO_2 is approximately 150 mmHg. At this relatively high level of P_IO_2, it is possible to achieve a normal P_AO_2 of 100 mmHg by alveolar ventilation. In comparison, at the peak of Mount Everest, because P_B is approximately 250 mmHg and P_IO_2 is only 43 mmHg, it is impossible to achieve a normal P_AO_2. Other adaptations of the respiratory and circulatory systems to high altitude are necessary for climbers to stay at the peak of Mount Everest for a brief time.

As shown in the previous scheme, in a steady state, the rate of oxygen supply by alveolar ventilation—difference in oxygen content between inspired air (F_IO_2) and alveolar air (F_AO_2)—equals the rate of metabolic O_2 consumption (VO_2), as shown in the following equation:

$$(F_IO_2 - F_AO_2) \times V_A = VO_2$$

F_IO_2 and F_AO_2 represent the volume of O_2 per volume of air in inspired air and alveolar air, respectively. PO_2 is proportional to FO_2 in both inspired and alveolar air. Accordingly, the above equation can be expressed in terms of P_IO_2 and P_AO_2:

$$(P_IO_2 - P_AO_2) \times V_A / K = VO_2$$

In the above equation, K is the proportional constant for the PO_2/FO_2 ratio in inspired and alveolar air. Rearranging this equation yields the following alveolar PO_2 equation:

$$\text{Alveolar } PO_2 \text{ Equation: } P_AO_2 = P_IO_2 - K \times VO_2/V_A$$

The alveolar PO_2 equation predicts that P_AO_2 is determined by three variables—P_IO_2, VO_2, and V_A. This equation is important for understanding the regulation of respiration in response to external perturbations—for example, exercise and high altitude. During exercise, VO_2 increases as a result of increase in metabolism. The alveolar PO_2 equation predicts that maintenance of a constant P_AO_2 during exercise requires a matching between increase in VO_2 by a proportional increase in V_A. At high altitude, P_IO_2 decreases as a result of a decrease in barometric pressure. The alveolar PO_2 equation predicts that maintenance of a constant P_AO_2 at high altitude requires a decrease in VO_2/V_A ratio, which can be achieved by decreasing VO_2 and/or increasing V_A. It is important to note that P_IO_2 is the upper limit for P_AO_2. That is, P_AO_2 cannot exceed P_IO_2 even when VO_2/V_A ratio is zero. Breathing 100 percent oxygen instead of air can be used to increase P_IO_2 at high altitude.

Alveolar Gas Equation. V_A is a critical determinant of both P_ACO_2 and P_AO_2, as shown in the alveolar PCO_2 and PO_2 equations. This characteristic implies that V_A-induced change in P_ACO_2 must be accompanied by opposite change in P_AO_2. For example, hypoventilation-induced increase in P_ACO_2 must be accompanied by a decrease in P_AO_2. Conversely, hyperventilation-induced decrease in P_ACO_2 must be accompanied by an increase in P_AO_2. It is possible to predict PAO_2 from $PACO_2$ by combining the alveolar PCO_2 and PO_2 equations into an alveolar gas equation.

$$\text{Alveolar } PCO_2 \text{ Equation: } P_ACO_2 = K \times VCO_2/V_A$$
$$\text{Alveolar } PO_2 \text{ Equation: } P_AO_2 = P_IO_2 - K \times VO_2/V_A$$

Substituting V_A from the P_ACO_2 equation into the P_AO_2 equation yields the following alveolar gas equation for relating P_ACO_2 and P_AO_2:

$$\text{Alveolar Gas Equation: } P_AO_2 = P_IO_2 - P_ACO_2 \times VO_2/VCO_2$$

In the above alveolar gas equation, VO_2/VCO_2 is the inverse of respiratory quotient (VCO_2/VO_2), which is a function of the substrate being utilized for metabolism. Respiratory quotient is 1 for carbohydrate metabolism and less than 1 for fat and protein metabolism. By assuming 0.8 as the average respiratory quotient, the alveolar gas equation can be simplified to the following equation:

$$P_AO_2 = P_IO_2 - 1.2 \times P_ACO_2$$

In the clinical setting, this simplified alveolar gas equation can be used to estimate P_AO_2 from P_IO_2 and P_ACO_2. P_IO_2 can be calculated from barometric pressure, inspired gas composition, and saturated water pressure. P_ACO_2 can be approximated by PCO_2 in end-expired air—air exhaled from the lungs at the end of expiration. For example, P_IO_2 for a patient breathing air at sea level is expected to be 150 mmHg. If the patient's P_ACO_2 is 40 mmHg, then P_AO_2 is expected to be 102 mmHg.

Ventilation/Perfusion Ratio is the ratio of ventilation of the alveolus with air versus perfusion of the capillaries surrounding the alveolus. There are millions of alveoli in the lungs. In an ideal situation, ventilation/perfusion ratio equals 1, when an alveolus is ventilated and the surrounding capillaries are perfused by blood, such that air equilibrates efficiently with blood for gas exchange. In one example of mismatching between ventilation and perfusion, ventilation/perfusion ratio equals zero, when an alveolus is not ventilated by air but the surrounding capillaries are perfused by blood. In another example of mismatching between ventilation and perfusion, ventilation/perfusion ratio equals infinity, when an alveolus is ventilated but the surrounding capillaries are not perfused by blood. In these two examples of mismatching between ventilation and perfusion, there is no gas exchange between air and blood. Heterogeneity in ventilation/perfusion ratios among alveoli is inefficient for gas exchange.

In normal lungs, two mechanisms facilitate the normalization of ventilation/perfusion ratio in alveoli. The first mechanism is low PCO_2-induced bronchoconstriction for decreasing ventilation in alveoli having abnormally high ventilation/perfusion ratio. This mechanism essentially distributes airflow to alveoli that receive blood perfusion. The second mechanism is hypoxic pulmonary vasoconstriction for decreasing perfusion of alveoli having an abnormally low ventilation/perfusion ratio. This mechanism essentially distributes blood flow to alveoli that receive ventilation. It is noteworthy that hypoxic vasoconstriction is a unique characteristic of pulmonary circulation. In contrast, hypoxia causes vasodilation in all organs perfused by the systemic circulation.

PCO_2-induced bronchoconstriction and hypoxic pulmonary vasoconstriction do not completely eliminate heterogeneity in ventilation/perfusion ratios among alveoli. Small heterogeneity in ventilation/perfusion ratios among alveoli results in a small difference between alveolar PO_2 and mixed

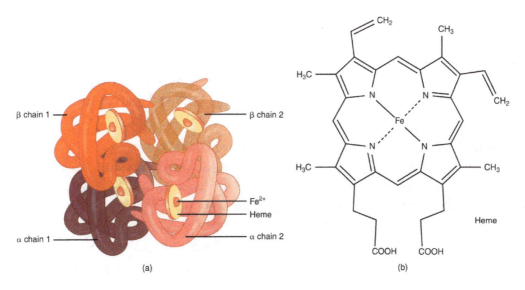

Fig. 7.10. Left Panel: Molecular structure of hemoglobin containing 2 α and 2 β protein subunits (brown and green ribbon models) and four heme groups (green stick models). Right Panel: Molecular structure of a heme group, of which the Fe²⁺ ion coordinates the binding of oxygen.

pulmonary venous PO_2, due to the nonlinearity between blood O_2 content and blood PO_2. In comparison, alveolar PCO_2 and mixed pulmonary venous blood PCO_2 are essentially identical, due to the linearity between blood CO_2 content and blood PCO_2. In some pulmonary diseases and during intense exercise, an increase in heterogeneity in ventilation/perfusion ratio can result in a large difference between alveolar PO_2 and mixed pulmonary venous PO_2.

BLOOD GAS TRANSPORT

Being lipid-soluble molecules, oxygen and carbon dioxide have relatively low solubility in plasma. Most of the oxygen in blood is carried by hemoglobin, and most of the carbon dioxide in blood is carried by bicarbonate. Dissolved oxygen and carbon dioxide in plasma represent minor fractions of the total oxygen and carbon dioxide in the blood.

Blood Oxygen Transport by Hemoglobin. As shown in Fig. 7.10, hemoglobin consists of four protein subunits and four ferrous (Fe^{2+})-containing heme groups for binding oxygen molecules. At 100 percent saturation, when all of the four heme groups of hemoglobin are bound by oxygen, the oxygen-binding capacity of hemoglobin is 1.36 ml O_2/g. Blood oxygen content ($[O_2]$) can be calculated from hemoglobin saturation (S), the oxygen-binding capacity of hemoglobin (1.36 ml/g), and hemoglobin concentration ($[Hb]$), as shown in the following equation:

$$\text{Blood } [O_2] = S \times 1.36 \times [Hb]$$

For systemic arterial blood at sea level, hemoglobin saturation is approximately 100 percent. Normal blood hemoglobin concentration is approximately 15 g/100 ml. Accordingly, oxygen content ($[O_2]_a$) of systemic arterial blood is 20.4 ml O_2/100 ml blood, as shown in the following calculation:

$$[O_2]_a = 100\% \times 1.36 \text{ ml } O_2/\text{g Hb} \times 15 \text{ g Hb}/100 \text{ ml} = 20.4 \text{ ml } O_2/100 \text{ ml}$$

Hemoglobin saturation and oxygen content are lower in mixed venous blood than systemic arterial blood due to the extraction of oxygen for consumption by organ systems. The arterial-venous difference in oxygen content is a function of oxygen consumption rate (VO_2) and cardiac output, as shown in the following equation, also known as the Fick principle:

$$\text{Fick Principle:} \quad VO_2 = \text{Cardiac Output} \times ([O_2]a - [O_2]v)$$

At rest, oxygen consumption rate is 250 ml O_2/min and resting cardiac output is 5,000 ml/min. According to the Fick principle, the mixed venous blood oxygen content has to be 15.4 ml O_2/100 ml, as shown in the following calculation:

$$[O_2]v = [O_2]a - VO_2/\text{Cardiac Output}$$
$$= 20.4 \text{ ml } O_2/100 \text{ ml} - 250 \text{ ml } O_2/\text{min}/5,000 \text{ ml}/\text{min} = 15.4 \text{ ml } O_2/100 \text{ ml}$$

Using the oxygen content equation, hemoglobin saturation in mixed venous blood has to be 75.5 percent, as shown in the following calculation:

$$S_v = [O_2]_v/(1.36 \times [Hb])$$
$$= 15.4 \text{ ml } O_2/100 \text{ ml} / (1.36 \text{ ml } O_2/\text{g Hb} \times 15\text{g}/100\text{ml}) = 75.5\%$$

The Fick principle highlights the importance of cardiac output and oxygen extraction in supporting oxygen consumption by organ systems. For example, the moderate increase in VO_2 during light exercise can be provided by an increase in cardiac output alone without significant change in arterial-venous difference in oxygen content. In comparison, maximum VO_2 during intense exercise typically requires increases in both cardiac output and arterial-venous oxygen extraction.

Anemia. This example illustrates the effect of anemia—low hemoglobin concentration—on oxygen consumption rate. To simulate anemia, assume that blood hemoglobin concentration decreases by 20 percent to 12 g/100 ml. If arterial and venous hemoglobin saturations remain the same, then arterio-venous difference in oxygen content will decrease by 20 percent to 4 ml O_2/100 ml, as shown in the following calculation:

$$[O_2]_a - [O_2]_v = (100\% - 75.5\%) \times 1.36 \text{ ml } O_2/\text{g Hb} \times 12 \text{ g Hb}/100 \text{ ml}$$

$$= 4 \text{ ml } O_2/100 \text{ ml}$$

If cardiac output remains the same at 5,000 ml/min, then the oxygen consumption rate will decrease by 20 percent to 200 ml/min, as shown in the following calculation:

$$VO_2 = 5,000 \text{ ml}/\text{min} \times 4 \text{ ml } O_2/100 \text{ ml} = 200 \text{ ml } O_2/\text{min}$$

This calculation implies that anemia can potentially result in the limitation of physical activity.

Anemia with Compensatory Increase Cardiac Output. This example illustrates that the potentially negative effect of anemia on VO_2 can be compensated by increasing cardiac output. As shown in the following calculation, maintaining VO_2 at 250 ml O_2/min requires increasing cardiac output by 25 percent to 6,250 ml/min:

$$\text{Cardiac Output} = VO_2 / ([O_2]_a - [O_2]_v) = 250 \text{ ml } O_2/\text{min} /(4 \text{ ml } O_2/100 \text{ ml})$$

$$= 6,250 \text{ ml}/\text{min}$$

Compensatory increase in cardiac output in response to moderate anemia is typically sufficient to maintain basal VO_2 at rest, but is often insufficient to support maximum oxygen consumption rate during intense exercise, when the heart reaches its maximum limit in producing cardiac output.

Hemoglobin is packaged inside red blood cells. Normal red blood cell volume/blood volume ratio—hematocrit—is approximately 45 percent in males and 40 percent in females. The synthesis of red blood cells is regulated by erythropoietin—a hormone produced by the kidney. Blood hemoglobin concentration is an important determinant of maximum VO_2 and exercise performance. Stimulation of red cell synthesis by the administration of erythropoietin is a type of illegal blood doping used by some athletes for boosting exercise performance.

Oxyhemoglobin Dissociation Curve. Each hemoglobin molecule binds up to four molecules of oxygen as a function of PO_2. The relationship between fractional saturation of hemoglobin with oxygen and PO_2 is known as the oxyhemoglobin dissociation curve, as shown in Fig. 7.11A. The oxyhemoglobin dissociation curve has a sigmoidal shape, because the binding affinity of hemoglobin for oxygen increases with sequential binding of oxygen to hemoglobin. The binding affinity of hemoglobin for the first oxygen molecule is relatively low, as indicated by the shallow slope of the oxyhemoglobin dissociation curve at low PO_2 (Fig. 7.11A). The binding affinity of hemoglobin for oxygen becomes higher after the first binding site of hemoglobin is occupied by oxygen, as indicated by the steeper slope of the hemoglobin dissociation curve at intermediate PO_2. The enhancing effect of binding of oxygen to one site on the binding affinity of the next site is known as "positive cooperativity."

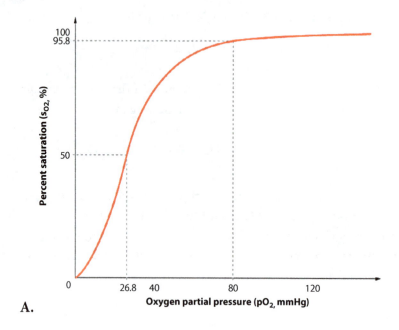

A.

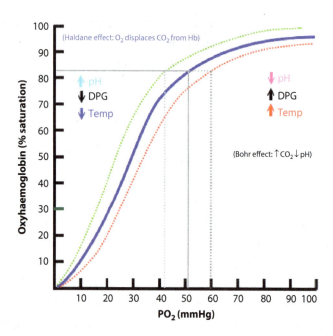

B.

Fig. 7.11—A. Oxyhemoglobin dissociation curve—showing a sigmoidal shape due to positive cooperativity in the binding of oxygen to the four subunits.   **B.** Shifting of oxyhemoglobin dissociation curve by pH, temperature, and DPG.

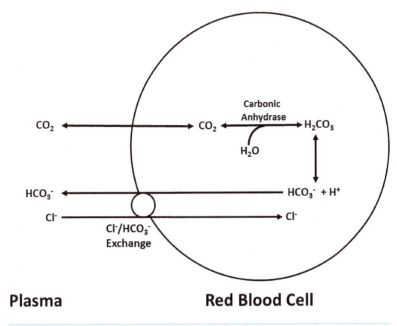

Fig. 7.12. Blood transport of carbon dioxide in the formation of bicarbonate.

The overall binding affinity of hemoglobin is inversely related to the half-maximal PO_2 (P_{50}) of the oxyhemoglobin dissociation curve—PO_2 at which 50 percent of O_2-binding sites are occupied. A low P_{50} indicates high binding affinity, and vice versa. As shown in Fig. 7.11A, P_{50} of hemoglobin in blood is approximately 27 mmHg. This moderately high P_{50} of hemoglobin in blood is caused by the negative regulation of hemoglobin's binding affinity by pH, 2,3-diphosphoglycerate (2,3-DPG), and temperature, as shown in Fig. 7.11B. In particular, 2,3-DPG—a metabolic intermediate of glycolysis—is important for maintaining the P_{50} of hemoglobin in blood within physiological range. During long-term blood storage, a time-dependent decrease in 2,3-DPG can result in abnormally low P_{50}—high binding affinity—of hemoglobin, which can hinder the unloading of oxygen to cells. During intense exercise, as a result of lactic acid and heat production, [H^+] and temperature increase in the vicinity of exercising muscle cells, which reduce the binding affinity of hemoglobin for oxygen and enhance the unloading of oxygen to muscle cells.

Blood Transport of Carbon Dioxide. Similar to oxygen, carbon dioxide has a relatively low solubility in plasma. Dissolved carbon dioxide represents only 10 percent of total carbon dioxide in arterial blood. Carbamino-hemoglobin accounts for approximately 30 percent of the total blood CO_2 content. Carbamino-hemoglobin is formed from a reaction between CO_2 and the amino termini of hemoglobin (Hb), as shown in the following reaction:

$$Hb\text{–}NH_2 + CO_2 \leftrightarrow Hb\text{–}NHCOOH \leftrightarrow Hb\text{–}NHCOO^- + H^+$$

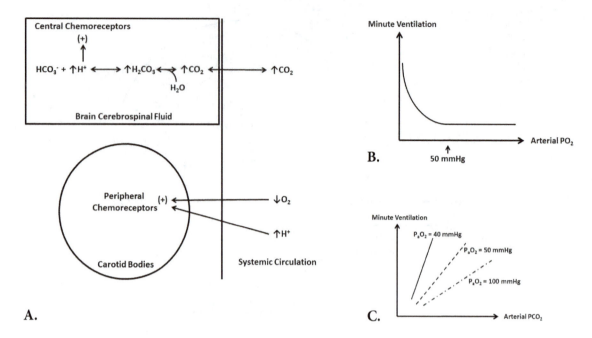

Fig. 7.13—A. Central and peripheral chemoreceptors for the regulation of respiration.   **B.** Dependence of minute ventilation on arterial PO$_2$. **C.** Dependence of minute ventilation on arterial PCO$_2$ at various PaO$_2$ levels ranging from 40 to 100 mmHg.

At physiological pH, most of carbamino hemoglobin is present in the charged form (Hb-NHCOO$^-$).

Majority of carbon dioxide ($\sim$60%) in the blood is transported in the form of bicarbonate (HCO$_3^-$), which is generated by the following chemical reaction:

$$CO_2 + H_2O \leftrightarrow H_2CO_3 \leftrightarrow H^+ + HCO_3^-$$

The rate-limiting step for bicarbonate production is the formation of carbonic acid (H$_2$CO$_3$) from CO$_2$ and H$_2$O, catalyzed by carbonic anhydrase. H$_2$CO$_3$ then dissociates spontaneously to form H$^+$ and HCO$_3^-$. The conversion of CO$_2$ to bicarbonate occurs inside red blood cells, because carbonic anhydrase is inside red blood cells. As shown in Fig. 7.12, bicarbonate is then transported from red blood cells to plasma via the HCO$_3$/Cl$^-$ exchanger. The buffering of H$^+$ produced from this chemical reaction by cellular and plasma proteins, including hemoglobin, facilitates the production of bicarbonate from CO$_2$.

REGULATION OF RESPIRATION

The respiratory center regulates ventilation in response to sensory input from peripheral and central chemoreceptors. **Central chemoreceptors** are situated in the brain, whereas **peripheral**

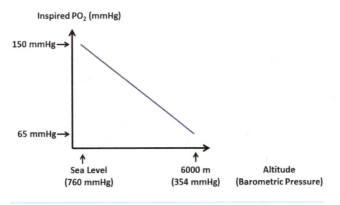

Fig. 7.14.   Dependence of inspired PO_2 on altitude.

chemoreceptors are located in aortic bodies at the aortic arch and carotid bodies at the carotid artery bifurcation. As illustrated in Fig. 7.13A, central chemoreceptors detect changes in cerebrospinal fluid $[H^+]$, which reflects blood PCO_2, whereas peripheral chemoreceptors detect changes in arterial blood PO_2 and $[H^+]$.

As shown in Fig. 7.13B, minute ventilation increases in response to severe hypoxia ($P_aO_2 <$ 50 mmHg), but is relatively insensitive to changes in PO_2 at levels higher than 50 mmHg. PO_2-sensing by peripheral chemoreceptors plays a critical role in stimulating ventilation at high altitude, when arterial PO_2 is near or below 50 mmHg. $[H^+]$-sensing by peripheral chemoreceptors plays an

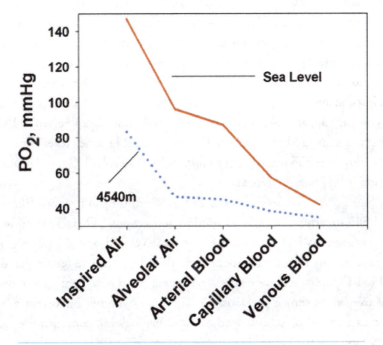

Fig. 7.15 PO_2 Levels from Inspired Air to Venous Blood at High Altitude and Sea Level (from Beall. PNAS 104 (Suppl 1): 8656, 2007).

important role in stimulating ventilation during extremely intense exercise, when lactic acid level increases in the plasma.

PCO_2-sensing by central chemoreceptors plays a critical role in the regulation of ventilation at sea level. As shown in Fig. 7.13C (**dotted dashed line**), at 100 mmHg P_aO_2, minute ventilation increases linearly with arterial PCO_2. As shown in Fig. 7.13C (**solid and dashed lines**), hypoxia (P_aO_2 < 50 mmHg) enhances the sensitivity of central chemoreceptors to P_aCO_2. The modulation of PCO_2-sensitivity of ventilation by hypoxia contributes to the hyperventilatory response of the respiratory system at extreme altitude.

HIGH ALTITUDE PHYSIOLOGY

The primary challenge of high altitude for survival is the decrease in inspired PO_2 with altitude, as a result of the decrease in barometric pressure. The following equation shows the inspired PO_2 for breathing air as a function of barometric pressure:

$$\text{Inspired PO}_2 = 0.21 \times (\text{Barometric Pressure} - 47\,\text{mmHg})$$

As shown in Fig. 7.14, at sea level, barometric pressure is 760 mm Hg, and inspired PO_2 is 150 mmHg,). With inspired PO_2 at 150 mmHg, it is feasible to maintain alveolar PO_2 at the normal level (100 mmHg) by a moderate level of alveolar ventilation for a reasonable level of physical activity. In comparison, as shown in Fig. 7.14, at 6,000 m—above the level for permanent human inhabitation—barometric pressure falls to 354 mm Hg, and inspired PO_2 falls to 65 mmHg. With inspired PO_2 at 65 mmHg, it is impossible to maintain alveolar PO_2 at 100 mmHg.

Fig. 7.15 shows adaptation of the respiratory system to high altitude by comparing profiles of PO_2—from inspired air to venous blood—in people living at 4,540 m and sea level. As shown in Fig. 7.15, at sea level, inspired PO_2 is approximately 150 mmHg and alveolar PO_2 is approximately 100 mmHg. In comparison, at 4,540 m, inspired PO_2 is 80 mmHg and alveolar PO_2 is 50 mmHg. The difference between inspired and alveolar PO_2 is 50 mmHg at sea level but only 30 mmHg at high altitude. The narrower difference between inspired and alveolar PO_2 at high altitude is achieved by hyperventilation in response to hypoxia.

As shown in Fig. 7.15, at sea level, arterial PO_2 is 90 mmHg and venous PO_2 is 40 mmHg. In comparison, at 4,540 m, arterial PO_2 is 45 mmHg and venous PO_2 is 30 mmHg. The difference between arterial and venous PO_2 is 50 mmHg at sea level and only 15 mmHg at high altitude. The smaller arterial-venous PO_2 difference at high altitude is feasible for oxygen delivery, because the 30–45 mmHg PO_2 range represents the steepest part of the oxyhemoglobin dissociation curve for unloading of oxygen, as shown previously in Fig. 7.11. Another important adaptation to high altitude is an increase in red blood cell synthesis and hemoglobin concentration, as stimulated by erythropoietin secretion by the kidney in response to hypoxia.

KEY TERMS

- airway resistance
- alveolar gas equation
- alveolar PCO$_2$ equation
- alveolar PO$_2$ equation
- alveolar pressure
- alveolar ventilation
- anemia
- atmospheric pressure
- barometric pressure
- blood gas transport
- central chemoreceptor

- dead space
- expiration
- flow-volume loop
- hemoglobin
- high altitude physiology
- inspiration
- intrapleural pressure
- Laplace's Law
- lung capacities
- lung compliance
- lung recoil

- lung volumes
- minute ventilation
- oxyhemoglobin dissociation curve
- peripheral chemoreceptor
- surface tension
- surfactant
- tissue elasticity
- transpulmonary pressure
- ventilation/perfusion ratio

IMAGE CREDITS

8

RENAL PHYSIOLOGY

Kidneys perform multiple functions in the human body—production of erythropoietin for stimulating red cell synthesis, activation of vitamin D by hydroxylation, production of glucose by gluconeogenesis, and regulation of blood volume, blood pressure, and extracellular fluid composition.

BASIC ANATOMY OF THE KIDNEY

As shown in Fig. 8.1A, there are two kidneys in the human body, each of which is connected to a renal artery, renal vein, and ureter. From the systems point of view, the primary function of kidneys is to remove waste products from the relatively dirtier renal arterial blood, excrete waste products to the urine, and return the relatively cleaner blood to the renal vein. As shown in Fig. 8.1A, urine from the two kidneys is drained via the two ureters into the urinary bladder for excretion through the urethra during urination. Fig. 8.1B shows a longitudinal section of a kidney (Fig. 8.1B, left panel) and a nephron—functional unit of kidney (Fig. 8.1B, right panel). A nephron consists of a renal vascular system interfaced

LEARNING OBJECTIVES

1. **Glomerular Filtration.** Explain the regulation of glomerular filtration by hydrostatic pressure and protein osmotic pressure in glomerular capillaries and Bowan's Capsule; compare and contrast afferent and efferent arterioles in terms of their function in the regulation of glomerular filtration rate and renal plasma flow.

2. **Renal Reabsorption and Renal Secretion.** Compare and contrast renal reabsorption and renal secretion in terms of mechanisms and function in renal excretion; describe the basolateral and luminal transport mechanisms in proximal tubular cells for the reabsorption of Na^+, phosphate, glucose, and water.

3. **Production of Concentrated and Dilute Urine.** Explain how countercurrent multiplication and antidiuretic hormone together regulate the production of concentrated and dilute urine.

4. **Regulation of Extracellular Fluid Volume.** Discuss the regulation of extracellular fluid volume and blood pressure by renin–angiotensin–aldosterone system.

5. **Renal Regulation of Extracellular pH.** Explain the buffering of extracellular pH by the CO_2/bicarbonate system using the Henderson–Hasselbalch equation, and discuss the three mechanisms of renal regulation of extracellular pH.

6. **Renal Clearance.** Define and provide examples of renal clearance for substances that are handled by glomerular filtration with or without reabsorption and secretion.

7. **Regulation of the Lower Urinary Tract.** Compare and contrast the continence phase and micturition phase of urinary bladder and lower urinary tract in terms of autonomic and somatic nervous control of detrusor smooth muscle, external urethral sphincter, and internal urethral sphincter.

8. **Dialysis for the Management of Chronic Kidney Failure.** Compare and contrast peritoneal dialysis and hemodialysis for managing kidney failure in terms of mechanisms and procedures.

Components of the urinary systems

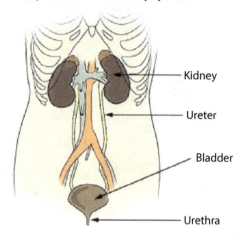

A.

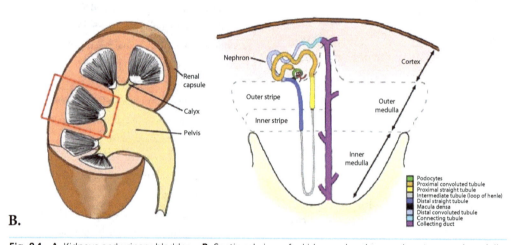

B.

Fig. 8.1—A. Kidneys and urinary bladder.   **B.** Sectional view of a kidney—showing renal cortex, renal medulla, and nephron.

with a renal tubular system. Fig. 8.1B **(right panel)** shows the beginning of the renal tubular system in the cortex, where it receives fluid of filtration from the vascular system. The renal tubular system then extends from the cortex into the inner medulla, loops back to the cortex, and finally extends back to the medulla for connection to the renal pelvis. Each segment of the renal tubular system performs a unique function.

NEPHRON—FUNCTIONAL UNIT OF THE KIDNEY

A nephron consists of a renal vascular system interfaced with a renal tubular system. Fig. 8.2A shows the interface between the renal vascular system and renal tubular system, consisting of glomerulus enveloped

(A)

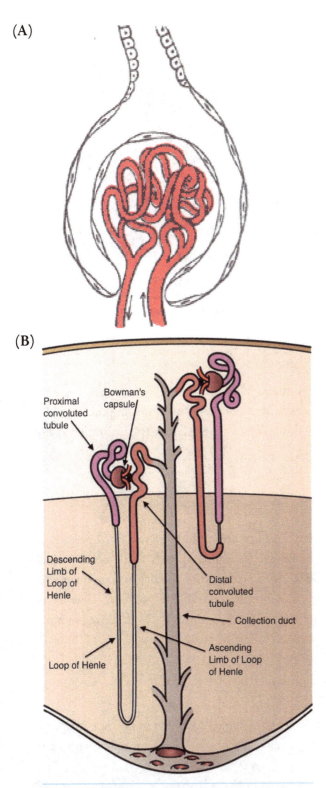

(B)

Proximal convoluted tubule

Bowman's capsule

Descending Limb of Loop of Henle

Loop of Henle

Distal convoluted tubule

Collection duct

Ascending Limb of Loop of Henle

Fig. 8.2—A. Structure of a renal corpuscle—glomerulus enveloped by Bowman's capsule. **B.** Structure of two nephrons—juxtaglomerular nephron having a long Loop of Henle and cortical nephron having short Loop of Henle .

by Bowman's capsule. Glomerulus is a network of porous capillaries having large pores in endothelial cells and large gaps between endothelial cells. Bowman's capsule is the beginning of the renal tubular system, consisting of epithelial cells with foot-like processes (podocytes) separated by large gaps. As a whole, the interface between glomerulus and Bowman's capsule is highly permeable to all small molecules other than proteins. Blood pressure in the glomerular capillaries drives rapid filtration of protein-free fluid from glomerular capillary into the Bowman's capsule. Glomerular filtrate is almost identical to plasma in composition, except for the absence of protein. Normal glomerular filtration rate is 125 ml/min, or 180 liters/day—i.e., 47 gallons/day. Typical urine flow is only 2 liters/day, because the vast majority (~99%) of glomerular filtrate is reabsorbed by the renal tubular system.

Fig. 8.2B shows the structure of juxtamedullary and cortical nephrons. A juxtamedullary nephron has a long loop of Henle that extends into the renal medulla, whereas a cortical nephron has a short loop of Henle that is situated mostly in the renal cortex. As shown in Fig. 8.2B, the glomerulus-Bowman's capsule filtration unit (also known as the renal corpuscle) forms the beginning of the renal tubular system. The next segment—proximal convoluted tubule—reabsorbs a substantial amount (60 percent) of glomerular filtrate without significantly changing the tubular fluid osmolarity, due to parallel reabsorption of solute and water by proximal tubular epithelial cells. The loop of Henle consists of a descending limb that extends from the renal cortex to the renal medulla, and an ascending limb that returns from the renal medulla to renal cortex. The primary function of the loop of Henle is production of an osmotic gradient in the renal interstitium from approximately 300 mOsM in the renal cortex to approximately 1,200 mOsM in the renal medulla. The juxtamedullary nephrons having a long loop of Henle are important for producing the osmotic gradient in the renal interstitium, whereas cortical nephrons without a long loop of Henle are relatively unimportant. As addressed later in this chapter, high osmolarity in the renal medulla is essential for the production of concentrated urine during dehydration.

The next segment—distal convoluted tubule—is known as the major diluting segment of the nephron, because distal tubular cells are capable of lowering tubular fluid osmolarity to lower than 100 mOsM by reabsorbing Na^+ from the tubular fluid to peritubular capillaries. Other diluting segments of the nephron include the ascending limb of the loop of Henle and the cortical collecting duct. The last segment of the nephron—medullary collecting duct—is important for producing concentrated urine during dehydration by allowing fluid in the collecting duct to equilibrate with the high osmolarity in renal medulla.

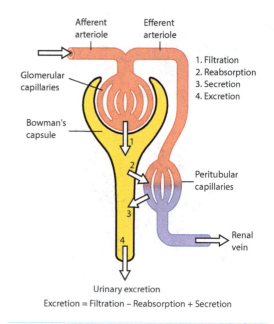

Fig. 8.3. Fundamental processes in the function of a nephron: 1. filtration, 2. reasbsorption, 3. secretion.

Fundamental Processes in a Nephron. Fig. 8.3 summarizes the three fundamental processes in a nephron: glomerular filtration, tubular reabsorption, and tubular secretion. Glomerular filtration is the first step of renal function.

GLOMERULAR FILTRATION

As shown in Fig. 8.3, blood enters the glomerulus from the renal artery. Approximately 20 percent of renal plasma flow enters the Bowman's capsule by glomerular filtration, whereas 80 percent of renal plasma flow enters peritubular capillaries. As cited previously, glomerular capillaries and Bowman's capsule epithelium are highly permeable to small molecules, but impermeable to proteins. Glomerular capillary blood pressure (P_{GC}) is higher than Bowman's capsule fluid pressure (P_{BC})—resulting in a net hydrostatic pressure for glomerular filtration, as shown in the following equation:

$$\text{Net Hydrostatic Pressure for Glomerular Filtration} = P_{GC} - P_{BC}$$

Protein osmotic pressure in glomerular capillary blood (π_{GC}) is higher than that in Bowman's capsule fluid (π_{BC}), resulting in a net protein osmotic pressure against glomerular filtration, as shown in the following equation:

Net Protein Osmotic Pressure Against Glomerular Filtration = $\pi_{GC} - \pi_{BC}$

Glomerular filtration pressure is the difference between net hydrostatic pressure for glomerular filtration and net protein osmotic pressure against glomerular filtration, as shown in the following equation:

Glomerular Filtration Pressure = $P_{GC} - P_{BC} - \pi_{GC} + \pi_{BC}$

Glomerular capillary blood pressure (P_{GC}) is normally the largest term in the above equation and the primary driving pressure for glomerular filtration. Glomerular capillary protein osmotic pressure (π_{GC}) is normally the second-largest term in the equation. Bowman's capsule fluid pressure (P_{BC}) and protein osmotic pressure (π_{BC}) are relatively small, because tubular fluid flows freely to the renal pelvis and proteins are normally not filtered into the Bowman's capsule.

Glomerular capillary blood pressure and glomerular filtration rate are dependent upon vascular resistances of afferent and efferent arterioles, because glomerular capillary is situated between afferent and efferent arterioles, as shown in Fig. 8.3. **Afferent arteriole** controls the blood pressure drop from the renal artery to the glomerular capillary, whereas the **efferent arteriole** controls the blood pressure drop from the glomerular capillary to the peritubular capillaries. Afferent arteriolar vasoconstriction decreases glomerular filtration pressure and glomerular filtration rate, because an increase in afferent arteriolar resistance increases the blood pressure drop from the renal artery to the glomerular capillary. In contrast, efferent arteriolar vasoconstriction increases glomerular filtration pressure and glomerular filtration rate, because an increase in efferent arteriolar resistance increases the pressure drop from the glomerular capillary to the peritubular capillary.

It is noteworthy that changes in **renal blood flow** (RBF) do not necessarily follow changes in glomerular filtration rate (GFR), because renal blood flow decreases in response to an increase in renal vascular resistance, regardless of the vessel of vasoconstriction, as summarized in the following table:

	GFR	RBF
Afferent Arteriolar Vasoconstriction :	Decrease	Decrease
Efferent Arteriolar Vasoconstriction:	Increase	Decrease

For example, afferent arteriolar vasoconstriction induced by sympathetic stimulation results in decreases in both glomerular filtration rate and renal blood flow. In comparison, efferent arteriolar vasoconstriction induced by angiotensin II results in an increase in the glomerular filtration rate but a decrease in renal blood flow.

(A)

(B)

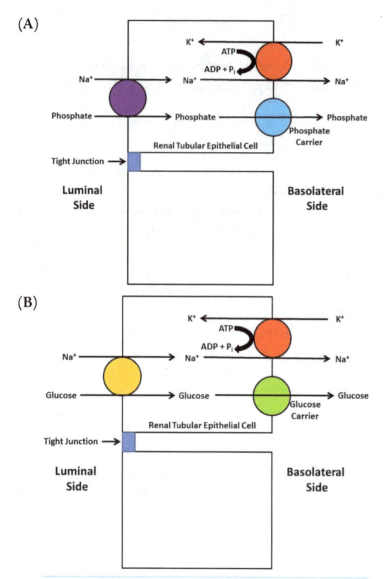

Fig. 8.4—Mechanisms of (A) Phosphate Reabsorption and (B) Glucose Reabsorption in the Proximal Tubule. Phosphate and glucose are transported from the tubular lumen into tubular cells by sodium-phosphate cotransport and sodium-glucose cotransport, respectively, resulting in high intracellular concentrations of phosphate and glucose. Phosphate and glucose are then transported from tubular cells to the interstitium by facilitated diffusion via carriers.

RENAL REABSORPTION AND SECRETION

As cited previously, approximately 20 percent of renal plasma flow enters the Bowman's capsule by glomerular filtration, whereas 80 percent of renal plasma flow enters peritubular capillaries. Renal reabsorption is the process of transporting a molecule by renal tubular cells from glomerular filtrate in the renal tubular system to peritubular capillaries, thereby decreasing the excretion of a molecule. Renal secretion is the process of transporting a molecule by renal tubular cells from peritubular capillaries into the renal tubular system, thereby increasing the excretion of a molecule.

Glomerular filtration is the common first step for the excretion of all molecules, but actual excretion of a molecule is also determined by the relative magnitudes of renal absorption and secretion, as shown in the following equation:

$$Excretion = Filtration - Reabsorption + Secretion$$

Excretion of creatinine, a product of protein metabolism, is determined by glomerular filtration only, because creatinine is neither reabsorbed nor secreted by renal tubular cells. Excretion of glucose is normally zero, because glucose is completely reabsorbed from glomerular filtrate in the tubular system to the peritubular capillaries. Excretion of para-aminohippuric acid (PAH), a diagnostic agent for renal function, is determined by renal plasma flow, because PAH enters the renal tubular system by glomerular filtration and is also secreted almost completely from peritubular capillaries into the renal tubular system.

Proximal Tubule—Major Site of Fluid Reabsorption. Proximal convolute tubule—the first renal tubular segment after Bowman's capsule—is a major site of renal reabsorption. More than 60 percent of glomerular filtrate is reabsorbed by the proximal tubule. As shown in Figs. 8.4, transporters on the luminal side of proximal tubular cells are Na^+-coupled transporters for driving reabsorption of molecules against their concentration gradient. Figs. 8.4A and 8.4B show renal tubular reabsorption of phosphate and glucose, respectively, by distinct transporters in the luminal and basolateral membranes of a proximal tubular cell. As shown in Fig. 8.4A, the Na^+-phosphate cotransporter in the luminal membrane of the proximal tubular cell transports phosphate from tubular fluid into the cell, thereby establishing a high concentration of phosphate inside the cell. The phosphate carrier in the basolateral membrane of the proximal tubular cell facilitates diffusion of phosphate down its concentration gradient from inside the cell to the interstitium, where phosphate diffuses into the peritubular capillary network. Na^+-K^+-ATPase in the basolateral membrane actively transports Na^+ from inside the cell into the interstitium, where Na^+ diffuses into the peritubular capillary network.

RENAL HANDLING OF GLUCOSE

As shown in Fig. 8.4B, the Na^+-glucose cotransporter, SGLT2, in the luminal membrane of the proximal tubular cell, transports glucose from tubular fluid into the cell, thereby establishing a high concentration of glucose inside the cell. The glucose carrier, GLUT2, in the basolateral membrane of the proximal tubular cell, facilitates diffusion of glucose down its concentration gradient from inside the cell to the interstitium, where glucose diffuses into the peritubular capillary network. Na^+ – K^+ – ATPase in the basolateral membrane actively transports Na^+ from inside the cell into the interstitium, where Na^+ diffuses into the peritubular capillary network.

In healthy subjects, glucose does not appear in the urine, because plasma [glucose] is below the renal threshold for complete reabsorption of glucose from glomerular filtrate. Renal threshold is the tubular concentration of glucose (11 mM) at which all binding sites of glucose transporters become saturated and renal tubular transport of glucose reaches the maximum transport rate (T_{max}). Normal plasma [glucose] at rest is typically 5 mM. Maximum plasma glucose concentration after a meal is typically lower than 9 mM, which is below the renal threshold for complete glucose reabsorption.

In patients having diabetes mellitus, plasma glucose concentration can exceed the renal threshold for complete glucose reabsorption, and some glucose will be excreted into the urine. As shown in the following equation, the rate of urinary glucose excretion equals the difference between T_{max} and the flux of glucose into the renal tubular system by glomerular filtration, which is a product of glomerular filtration rate (GFR) and plasma glucose concentration ($[glucose]_{plasma}$):

$$\text{Rate of Urinary Glucose Excretion} = GFR \times [Glucose]_{plasma} - T_{max}$$

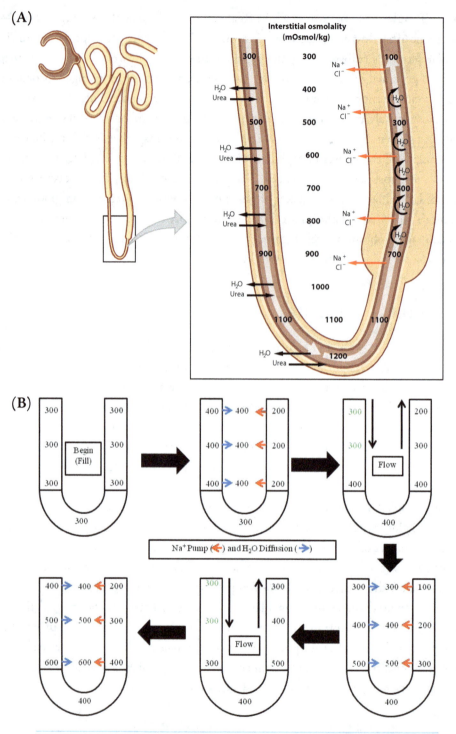

Fig. 8.5—A. Creation of the concentration gradient from cortex to medulla by differential locations of sodium, urea, and water transporters in the nephron.   **B.** Model of counter-current multiplication.

High concentration of glucose is toxic to organ systems. To lower the plasma glucose in patients having type 2 diabetes mellitus, pharmacologic inhibitors of the Na$^+$-glucose cotransporter, SGLT2, have been developed for blocking renal reabsorption of glucose, thereby increasing the excretion of glucose.

Renal excretion of H$^+$ by Na$^+$/H$^+$ exchange and H$^+$-ATPase in the luminal membrane of proximal tubular cells will be covered later in the section on acid–base regulation. The high water permeability of the proximal tubule allows the reabsorption of water to follow the reabsorption of Na$^+$ from tubular fluid to the peritubular capillary network.

The osmolarity of fluid leaving the proximal tubule remains isoosmotic relative to the osmolarity of plasma, because the high water permeability of proximal tubule allows water reabsorption to follow solute reabsorption. Fluid reabsorption by the proximal tubule is known as isoosmotic reabsorption.

PRODUCTION OF CONCENTRATED URINE

Loop of Henle—Site of Countercurrent Multiplication for Generating Osmotic Gradient in the Renal Interstitium. As shown in Fig. 8.5A, the loop of Henle—tubular segment following proximal convoluted tubule—extends from the cortex to the medulla, and then loops back to the cortex. A major function of the loop of Henle is the generation of an osmotic gradient in the renal interstitium by countercurrent multiplication. "Countercurrent" refers to the opposite direction of fluid flow in the descending and ascending limbs of the loop of Henle. "Multiplication" refers to amplification of the Na$^+$ pumping effect of each segment of the loop to generate a large overall osmotic gradient in the renal interstitium. Each segment of the loop is capable of generating approximately 200 mOsM gradient between the renal tubular fluid and renal interstitium, but the overall osmotic gradient in the renal interstitium ranges from 300 mOsM in the renal cortex to 1,200 mOsM in the renal medulla, as shown in Fig. 8.5A.

As shown in Fig. 8.5A, the descending limb of the loop of Henle is enriched in aquaporin channels for high water permeability, whereas the ascending limb is enriched in Na$^+$ pumping activity. The high Na$^+$ pumping activity in the ascending limb creates a local osmotic gradient that is approximately 200 mOsM higher in the renal interstitium than tubular fluid in the ascending limb. The high water permeability of the descending limb causes tubular fluid in the descending limb to reach the osmolarity of the renal interstitium. The countercurrent fluid flow transfers fluid of high osmolarity from the descending limb to the ascending limb for further concentration.

The six panels in Fig. 8.5B illustrate the concept of countercurrent multiplication by simulating countercurrent flow and Na$^+$ pumping activity/water diffusion as separate steps. Fig. 8.5B (**upper left panel**) shows the initial condition for the simulation—filling of the loop of Henle by isoosmotic fluid (300 mOsM) from the proximal tubule. Fig. 8.5B (**upper middle panel**) simulates the effect of Na$^+$ pumping activity of the ascending limb in generating a local osmotic gradient—200 mOsM inside the ascending limb and 400 mOsM in the renal interstitium. Fluid osmolarity inside the descending limb is also 400 mOsM, because the high water permeability of the descending limb enables equilibration of tubular fluid with renal interstitium. Fluid osmolarity at the bottom of the loop remains at 300 mOsM due to the lack of Na$^+$ pumping activity there.

Fig. 8.5B **(upper right panel)** simulates the effect of countercurrent flow on moving fluid from the descending limb into the ascending limb. In this simulation, some isoosmotic fluid (300 OsM) enters the descending limb and pushes some fluid from the descending limb into the ascending limb, resulting in an osmotic gradient in the ascending limb ranging from 200 mOsM at the cortex to 400 mOsM at the renal medulla. Fig. 8.5B **(lower right panel)** simulates the effect of Na^+ pumping activity on creating a local osmotic gradient along the ascending limb, and equilibration of osmolarity between fluid in the descending limb and the renal interstitium. Na^+ pumping activity creates a local osmotic gradient of 200 mOsM in each segment of the ascending limb. Fluid osmolarity in the ascending limb becomes 100 mOsM at the renal cortex and 300 mOsM at the renal medulla, whereas osmolarity in the renal interstitium ranges from 300 mOsM in the renal cortex to 500 mOsM in the renal medulla. Similarly, fluid osmolarity in the descending limb becomes 300 mOsM at the renal cortex and 500 mOsM at the renal medulla, because the high water permeability of the descending limb enables equilibration of tubular fluid with renal interstitium. Altogether, osmolarity in the renal interstitium now ranges from 300 mOsM at the renal cortex to 500 mOsM at the renal medulla.

Fig. 8.5B **(lower middle panel)** simulates the effect of countercurrent flow on moving fluid from the descending limb into the ascending limb. In this simulation, some isoosmotic fluid (300 mOsM) enters the descending limb and pushes some fluid from the descending limb into the ascending limb, resulting in an osmotic gradient in the fluid inside the ascending limb—ranging from 300 mOsM at the cortex to 500 mOsM at the renal medulla. Fig. 8.5B **(lower left panel)** simulates the effect of Na^+ pumping along the ascending limb in generating new osmotic gradients along the ascending limb. Fluid osmolarity inside the ascending limb becomes 200 mOsM at the cortex and 400 mOsM at the renal medulla, whereas osmolarity in the renal interstitium becomes 400 mOsM in the cortex and 600 mOsM in the renal medulla. As illustrated in Fig. 8.5, Na^+ pumping along the ascending limb, high water permeability of the descending limb, and countercurrent flow together creates the high osmolarity in the renal medulla.

Urea, a product of protein metabolism, also contributes to the high osmolarity in the renal medulla. Urea accumulates in the renal medulla by two mechanisms. First, as a result of water reabsorption, urea becomes concentrated passively in the tubular fluid along the renal tubular system. Second, as shown in Fig. 8.5A, urea leaves the renal tubular system at the medullary collecting duct and reenters bottom portion of the loop of Henle, because urea transporters are enriched at these locations. Urea is transported by facilitated diffusion from the medullary collecting duct to the renal medulla, causing the concentration of urea in the renal medulla. The high concentration of urea in the renal medulla then drives the transport of urea by facilitate diffusion into the bottom portion of the loop of Henle.

Regulation of Extracellular Fluid Osmolarity. Maintenance of normal plasma osmolarity is essential for survival, because cell volume is determined mostly by extracellular fluid osmolarity. Abnormally low extracellular fluid osmolarity causes cell swelling, whereas abnormally high extracellular fluid osmolarity causes cell shrinking. Cell swelling is especially damaging to the brain, which has limited space for expansion inside the skull.

Extracellular fluid osmolarity is regulated by a negative feedback system, consisting of osmoreceptors in the hypothalamus for sensing extracellular fluid osmolarity, hypothalamic neuroendocrine cells for synthesizing **antidiuretic hormone** (ADH, also known as **vasopressin** (VP)) and releasing the hormone

A.   Urine Osmolarity (mOsM)

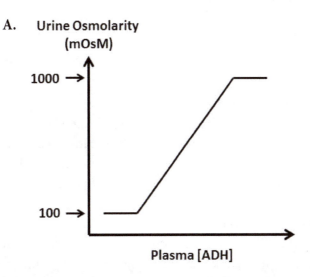

B.

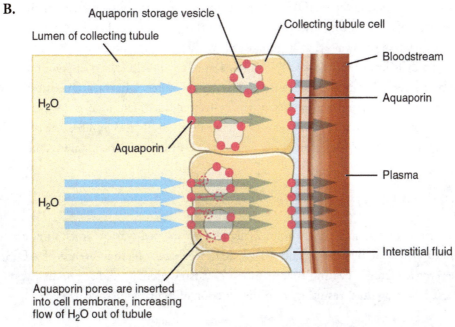

Fig. 8.6—A. Effect of antidiuretic hormone (vasopressin, VP) on urine osmolarity.   **B.** ADH/VP-dependent regulation of water permeability in renal collecting duct cells.

in the posterior pituitary gland, and the effector organ—kidney—for regulating water reabsorption. In response to high extracellular fluid (ECF) osmolarity, hypothalamic neuroendocrine cells release vasopressin into the circulation, which carries vasopressin to the kidneys, where vasopressin causes the kidney to increase water absorption and urine osmolarity, thereby conserving water for restoring extracellular fluid osmolarity, as summarized in the following scheme:

$$\text{High ECF Osmolarity} \rightarrow \uparrow \text{Plasma ADH/VP} \rightarrow \uparrow \text{Renal Reabsorption of Water}$$

Fig. 8.6A shows the effect of ADH/VP on increasing urine osmolarity from as low as 100 mOsM in the absence of ADH/VP to 1,000 mOsM in the presence of ADH/VP. In addition to regulating renal water reabsorption, ADH/VP stimulates thirst for drinking water—a behavioral response for restoring normal plasma osmolarity. Patients having genetic deficiency in ADH/VP expression are not capable of producing concentrated urine. These patients tend to produce a large amount of urine having low osmolarity—a condition known as diabetes insipidus.

ADH/VP increases renal reabsorption of water by increasing the water permeability of the collecting duct, thereby allowing the equilibration of fluid osmolarity in the collecting duct with the high osmolarity in the renal medulla. Fig. 8.6B shows the cellular mechanism by which ADH/VP increases water permeability of the collecting duct. ADH/VP binds to a G protein-coupled receptor that is coupled by G_s to the activation of adenylate cyclase, which catalyzes the production of cyclic AMP from ATP. Cyclic AMP activates signaling pathways that stimulate the synthesis of aquaporin water channels, the package of aquaporin channels in membrane vesicles, and fusion of aquaporin channels-containing vesicles with the cell membrane, resulting in the insertion of aquaporin channels into the cell membrane of collecting duct cells for increasing water permeability of the collecting duct. The high water permeability of the collecting duct enables production of concentrated urine by allowing the equilibration of fluid osmolarity in the collecting duct with the high osmolarity (up to 1,200 mOsM) in the renal medulla.

PRODUCTION OF DILUTE URINE

Excess consumption of water can cause abnormally low plasma osmolarity, which inhibits the secretion of ADH by the posterior pituitary gland via negative feedback. In the absence of ADH, aquaporin channels are removed from collecting duct membrane, causing tubular fluid to remain hypoosmotic as it passes through collecting duct, resulting in the production of dilute urine.

REGULATION OF EXTRACELLULAR FLUID VOLUME

Maintenance of extracellular fluid volume is essential for survival, because plasma volume, as part of the extracellular fluid volume, determines the extent of ventricular filling and stroke volume of the heart, as predicted by the Frank-Starling mechanism of the heart. During severe hemorrhage and dehydration, an extreme decrease in extracellular fluid volume can lead to loss of stroke volume, loss of cardiac output, and

A.

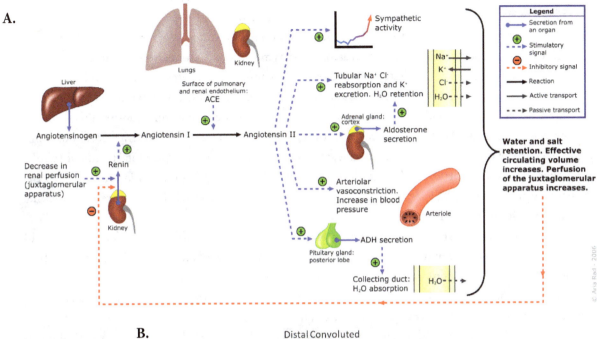

B.

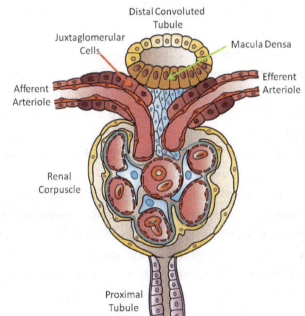

Fig. 8.7—A. Renin-angiotensin-aldosterone system for the regulation of blood volume and blood pressure. **B.** Juxtaglomerular apparatus—consisting of macula densa in the convoluted tubule and juxtaglomerular cells in the afferent arteriole.

death. Extracellular fluid (ECF) volume is regulated largely by the renin-angiotensin-aldosterone system, a cascade of enzymatic reactions carried out by multiple organ systems—liver, kidney, lung, and adrenal gland. As shown in Fig. 8.7A, in response to a decrease in renal perfusion as the result of a decrease in ECF volume, the kidney secretes renin, a proteolytic enzyme, into the circulation. Renin catalyzes the conversion of angiotensinogen, a circulating precursor 485 amino acid protein synthesized by the liver, to angiotensin I, a relatively inactive 10 amino acid peptide. Angiotensin I is converted to angiotensin II, a highly active 8 amino acid peptide, by angiotensin-converting enzymes on endothelial cells in the pulmonary and renal circulation. Angiotensin II stimulates the release of aldosterone, a steroid hormone, by adrenal cortical cells to the circulation. Aldosterone enhances Na^+ reabsorption in the **distal tubule**, connecting tubule, and cortical collecting duct by stimulating the expression of Na^+ channels on the luminal side and Na^+-K^+-ATPase on the basolateral side of tubular epithelial cells, thereby conserving Na^+ and water for restoring normal plasma volume.

Fig. 8.7B shows the structure of the **juxtaglomerular apparatus** in the kidneys for the release of renin to the circulation. The juxtaglomerular apparatus consists of **macula densa** cells in the distal tubule in close contact with juxtaglomerular cells in the afferent arteriole. Macula densa cells are sensors of Na^+ delivery through distal tubule. In response to a low rate of Na^+ delivery through the distal tubule, macula densa cells signal juxtaglomerular cells to release renin.

In addition to stimulating aldosterone release, angiotensin II exerts multiple effects on several organ systems—adrenal cortex, vascular system, sympathetic nervous system, kidney, and posterior pituitary gland. As shown in Fig. 8.7A, angiotensin II stimulates arteriolar vasoconstriction in the systemic circulation, thereby increasing total peripheral resistance and arterial blood pressure in the systemic circulation. Angiotensin II stimulates efferent arteriolar vasoconstriction in the kidney, thereby causing an increase in the glomerular filtration rate. Angiotensin II also stimulates sympathetic activity in the central nervous system, which can lead to increases in cardiac output and blood pressure. Angiotensin II stimulates the secretion of ADH by the posterior pituitary gland, which can lead to an increase in renal water absorption, thereby increasing plasma volume.

As shown in Fig. 8.7A (**red dotted line**), the renin-angiotensin-aldosterone system functions as negative feedback for maintaining normal blood volume and blood pressure. Over-activity of the renin-angiotensin-aldosterone system can cause an abnormal increase in blood volume and hypertension in some patients. Inhibitors of the renin-angiotensin-aldosterone system—for example, angiotensin-converting enzyme inhibitor, angiotensin II, and aldosterone receptor antagonist—are available for the treatment of hypertension.

RENAL REGULATION OF EXTRACELLULAR FLUID CA^{2+}

Maintenance of normal extracellular fluid calcium concentration is essential for the function of all organ systems. For example, cardiac muscle contraction and neurotransmitter release are dependent on Ca^{2+} influx from the extracellular fluid into the cytoplasm through calcium channels on the cell membrane. Plasma Ca^{2+} is filtered into the renal tubular system by glomerular filtration. More than 80 percent of

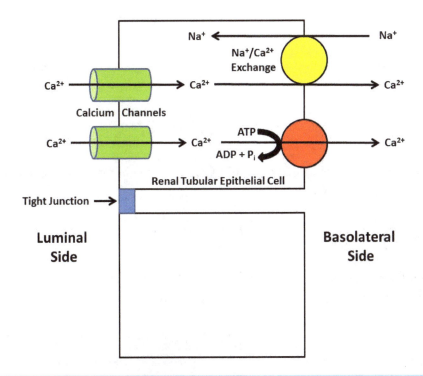

Fig. 8.8—Calcium reabsorption by distal tubular cells in the kidney. Ca²⁺ from the luminal side enters the tubular cell via a calcium channel, and then extruded out of the basolateral side via Na⁺/Ca²⁺ exchanger and Ca²⁺-ATPase.

filtered Ca^{2+} is reabsorbed passively in the proximal tubule and ascending limb of the loop of Henle through paracellular pathways, as driven by Ca^{2+} concentration gradient produced by Na^+ and water reabsorption.

Approximately 15–20 percent of filtered Ca^{2+} is reabsorbed by tubular cells in the distal tubule and connecting tubule under the regulation by parathyroid hormone and vitamin D. Fig. 8.8 shows the cellular mechanism of Ca^{2+} reabsorption by tubular cells. Ca^{2+} enters the luminal side of a tubular cell through calcium channels, and leaves the basolateral side through Na^+/Ca^{2+} exchange and Ca^{2+}-ATPase. Parathyroid hormone, vitamin D, and estrogen enhance renal Ca^{2+}-reabsorption by stimulating protein expression of Ca^{2+} transporters in tubular epithelial cells.

ACID–BASE REGULATION

The maintenance of normal extracellular pH within a narrow range—6.8 to 7.7—is essential for survival. Normal extracellular pH is maintained at 7.4 by respiratory and renal control of the following CO_2/bicarbonate buffer system:

$$CO_2 + H_2O \leftrightarrow H_2CO_3 \leftrightarrow H^+ + HCO_3^-$$

The following equation describes the chemical equilibrium of this buffer system, where K is the equilibrium constant for dissociation:

$$K = [H^+] \times [HCO_3^-] / [H_2CO_3]$$

The above equation can be expressed in terms of PCO_2, as shown in the following equation, where α is constant relating $[H_2CO_3]$ to PCO_2:

$$K = [H^+] \times [HCO_3^-] / \alpha PCO_2$$

The logarithmic form of the above equation is known as the Henderson-Hasselbalch equation, where pH and pK represent negative logs of $[H^+]$ and $[K]$, respectively.

$$pH = pK + \log\{[HCO_3^-]/\alpha PCO_2\}$$

The Henderson-Hasselbalch equation predicts that extracellular pH is determined by the $[HCO_3^-]/PCO_2$ ratio. From a simplistic point of view, the respiratory system regulates extracellular pH by regulating PCO_2 in the extracellular fluid, and the kidneys regulate extracellular pH by regulating $[HCO_3^-]$ in the extracellular fluid.

In principle, HCO_3^- can be lost from the extracellular fluid by glomerular filtration of HCO_3^- from plasma into the renal tubular system and neutralization of HCO_3^- by H^+ produced from protein and fat metabolism. The kidneys capture HCO_3^- in the glomerular filtrate completely by tubular reabsorption. $[HCO_3^-]$ in the urine is normally zero. The kidneys excrete H^+ by tubular secretion of H^+ and ammonium (NH_4^+).

Renal Reabsorption of Bicarbonate. Bicarbonate is filtered from the plasma into the renal tubular system by glomerular filtration, but is completely reabsorbed back to the circulation by renal tubular cells. Approximately 80 percent of filtered HCO_3^- is reabsorbed in the proximal tubule, while the remaining 20 percent of filtered HCO_3^- is reabsorbed in the ascending limb of the loop of Henle, distal convoluted tubule, and collecting duct. As shown in Fig. 8.9A, tubular epithelial cells reabsorb filtered HCO_3^- indirectly. Inside a tubular cell, carbonic anhydrase catalyzes the conversion of CO_2 and H_2O to H_2CO_3, which dissociates to H^+ and HCO_3^-. H^+ is transported from a tubular cell into tubular lumen by Na^+/H^+ exchange and H^+-ATPase at the luminal membrane for neutralization of luminal HCO_3^- to form H_2CO_3, which is converted to CO_2 and H_2O by carbonic anhydrase situated on the luminal membrane. Luminal CO_2 can enter a tubular cell and blood by simple diffusion. HCO_3^- is transported from a tubular cell into extracellular fluid and circulation via Na^+-HCO_3^- cotransport and HCO_3^-/Cl^- exchange at the basolateral membrane.

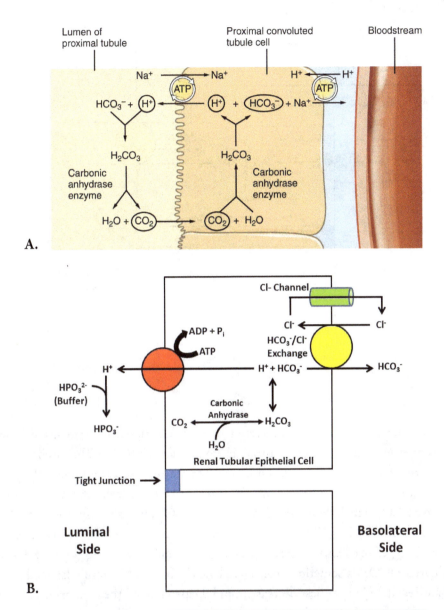

Fig. 8.9. Cellular mechanisms of **A.** Renal reabsorption of bicarbonate and **B.** Excretion of "titratable" acid by renal tubular cells.

As shown in Fig. 8.9A, carbonic anhydrase plays a critical role in renal HCO_3^- reabsorption. Inhibitors of carbonic anhydrase are sometimes used to increase HCO_3^- excretion for the treatment of respiratory alkalosis caused by hyperventilation at high altitude. Hyperventilation is necessary for blood oxygenation at high altitude, but alkalosis suppresses the respiratory center. Inhibition of carbonic anhydrase enhances HCO_3^- excretion and decreases plasma $[HCO_3^-]$, thereby normalizing $[HCO_3^-]/PCO_2$ ratio and extracellular pH.

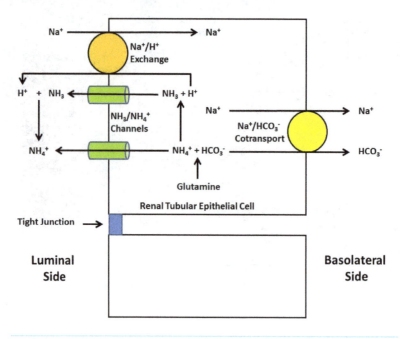

Fig. 8.10. Excretion of ammonium ion by renal tubular cells.

Renal Excretion of Titratable Acid. Renal excretion of H^+ production from protein and fat metabolism is necessary for preventing the H^+ from neutralizing extracellular HCO_3^-. The mechanism for H^+ excretion is similar to that for HCO_3^- reabsorption, except that the excreted H^+ is buffered by phosphate in the urine. As shown in Fig. 8.9B, carbonic anhydrase inside a tubular cell catalyzes the conversion of CO_2 with H_2O to H_2CO_3, which dissociates into H^+ and HCO_3^-. H^+ is transported actively from a tubular cell into luminal fluid by H^+-ATPase at the luminal membrane, and HCO_3- is transported to the circulation via HCO_3^-/Cl^- exchange at the basolateral membrane. The HCO_3^- generated by tubular cells in excess of filtered HCO_3^- is sometimes considered "new" HCO_3^-. H^+ in tubular fluid is buffered mostly by phosphate, resulting in a modest decrease in pH. Depending on the diet, urine pH can range from 5 to 8. A person eating a Western diet—high in meat products and low in vegetables—typically produces acidic urine. The amount of H^+ captured by buffer in the urine is known as "titratable acid," because it can be determined by titrating the pH of urine back to 7.4.

Renal Excretion of Ammonium. Renal excretion of ammonium (NH_4^+) contributes to almost 50 percent of total renal excretion of H^+. To calculate total amount of renal acid excretion, it is necessary to measure total ammonia in addition to titratable acid in the urine.

As shown in Fig. 8.10, renal tubular cells produce HCO_3^- and NH_4^+ from glutamine. HCO_3- is transported from a tubular cell to the circulation via Na^+–HCO_3^- cotransport in the basolateral membrane. NH_4^+ is excreted into tubular fluid through channels and Na^+/NH_4^+ exchange at the luminal membrane.

Renal Clearance

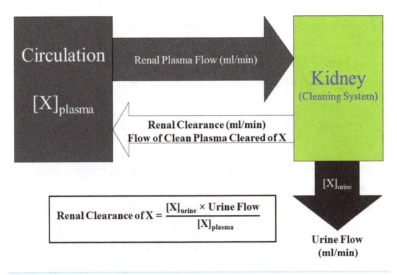

Fig. 8.11. Renal clearance is the flow of clean plasma cleared of X.

NH_4^+ is in equilibrium with ammonia (NH_3) and H^+ inside tubular cells and in tubular fluid, as shown in the following reaction:

$$NH_4^+ \leftrightarrow NH_3 + H^+$$

At pH 7.4, more than 98 percent of total ammonia exists as NH_4^+, whereas less than 2 percent of total ammonia exists as NH_3. As shown in Fig. 8.10, inside a tubular cell, NH_4^+ is in equilibrium with H^+ and NH_3. H^+ is transported from a tubular cell into tubular fluid via Na^+/H^+ exchange on the luminal side, and NH_3 is transported from a tubular cell into tubular fluid via channels on the luminal side.

RENAL CLEARANCE

Renal Clearance is a measure of the kidney's rate of clearing plasma of a given molecule X by urinary excretion, as illustrated in Fig. 8.11. Quantitatively, renal clearance is the flow of clean plasma cleared of X, as defined by the following equation:

$$\text{Clearance of X} = [X]_{urine} \times \text{Urine Flow} / [X]_{plasma}$$

$[X]_{urine}$ and $[X]_{plasma}$ represent the concentrations of X in urine and plasma, respectively. The units for urine flow and clearance are both volume/time. The equation for clearance is often remembered as: C = UV/P, where C is clearance for substance X, U is urine concentration of X, V is urine flow, and P is plasma concentration of X.

The product: $[X]_{urine}$ × Urine Flow represents rate of urinary excretion of X in mole per min, and the clearance equation can be expressed in the following alternate form:

$$\text{Clearance of X} = \text{Rate of Urinary Excretion of X} / [X]_{plasma}$$

To understand the concept of renal clearance, it is helpful to consider some specific examples of renal clearance.

Substances That Are Filtered Only—Clearance Equals Glomerular Filtration Rate (GFR). Creatinine, a product of muscle metabolism, is filtered into the renal tubular system by glomerular filtration, but is neither reabsorbed nor secreted by tubular cells. The rate of urinary excretion of creatinine equals the rate of entry of creatinine into the renal tubular system by glomerular filtration, as shown in the following equation:

$$\text{Rate of Urinary Excretion of Creatinine} = [Creatinine]_{plasma} \times \text{GFR}$$

As cited previously, clearance of a substance can be calculated by the following equation:

$$\text{Clearance of X} = \text{Rate of Urinary Excretion of X} / [X]_{plasma}$$

Substituting the creatinine excretion equation into the clearance equation yields the following equation:

$$\text{Clearance of Creatinine} = \text{GFR}$$

This equality can also be understood by recognizing that the glomerular filtration rate is the flow of plasma into the renal tubular system that can be cleared of creatinine. Estimated glomerular filtration rate based on creatinine clearance is an important indicator of kidney function and an important criterion for staging kidney diseases.

In practice, clearance of creatinine can be calculated by collecting blood for measuring $[creatinine]_{plasma}$ and collecting urine typically over twenty-four hours for measuring urine flow and $[creatinine]_{urine}$.

Inulin, a plant-derived polysaccharide, is also handled by glomerular filtration only, and clearance of inulin also equals the glomerular filtration rate. The procedure for measuring inulin clearance in a patient consists of infusing inulin into the circulation until plasma inulin concentration has reached steady state. Blood samples are then collected for measuring plasma inulin concentration, and urine samples are collected for measuring rate of urinary inulin excretion. Urine collection is sometimes omitted, because the rate of inulin infusion should equal rate of inulin excretion in a steady state.

Substances that are Filtered and Reabsorbed—Clearance is Lower Than the Glomerular Filtration Rate (GFR). When a substance X is filtered into the renal tubular system by glomerular filtration, and then reabsorbed by tubular cells, the rate of urinary excretion is less than the rate of delivery of the substance carried by glomerular filtration, as shown in the following relation:

$$\text{Rate of Urinary Excretion of X} < [X]_{plasma} \times \text{GFR}$$

As cited previously, clearance of a substance can be calculated by the following equation:

$$\text{Clearance of X} = \text{Rate of Urinary Excretion of X} / [X]_{plasma}$$

Substituting rate of urinary excretion of X from the relation into the clearance equation yields the following relation:

$$\text{Clearance of X} < \text{GFR}$$

For example, most of filtered Na^+ is reabsorbed by the renal tubular system, and clearance for Na^+ is typically below the glomerular filtration rate. Glucose and HCO_3^- are filtered into the renal tubular system by glomerular filtration, but are normally completely reabsorbed by tubular cells. Clearance of glucose and HCO_3^- are normally zero, because urinary excretions of glucose and HCO_3^- are normally zero.

Substances that are Filtered and Secreted—Clearance is Higher Than the Glomerular Filtration Rate (GFR). As cited previously, 20 percent of renal plasma flow is filtered into the renal tubular system by glomerular filtration. The remaining 80 percent of renal plasma flow enters the peritubular capillaries, where substances can be secreted into the renal tubular system by tubular cells for urinary excretion. When a substance X is filtered into the renal tubular system by glomerular filtration, and also secreted by tubular cells, the rate of urinary excretion is higher than the rate of delivery of the substance carried by glomerular filtration, as shown in the following relation:

$$\text{Rate of Urinary Excretion of X} > [X]_{plasma} \times \text{GFR}$$

As cited previously, clearance of a substance can be calculated by the following equation:

$$\text{Clearance of X} = \text{Rate of Urinary Excretion of X} / [X]_{plasma}$$

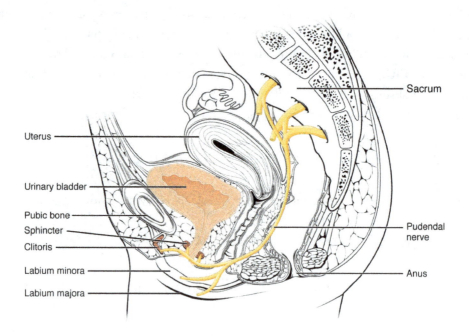

Fig. 8.12—Basic Neural Control of Bladder Function. The somatic pudendal nerve promotes bladder continence by stimulation the contraction of external sphincter. Not shown in this figure, the sympathetic hypogastric nerve exit from the thoracolumbar segment of the spinal cord also promotes bladder continence by stimulating the contraction of internal sphincter, whereas The parasympathetic pelvic nerve exit from the sacral region of the spinal promotes bladder emptying by stimulating the contraction of detrusor smooth muscle.

Substituting rate of urinary excretion of X from the relation into the clearance equation yields the following relation:

$$\text{Clearance of } X > GFR$$

For example, para-aminohippuric acid (PAH), a diagnostic agent for assessing renal perfusion, is almost completely secreted from peritubular capillaries into the tubular system for urinary excretion. Clearance of PAH is higher than GFR. Clearance of PAH approximates renal plasma flow, because the kidneys clear PAH almost completely from the plasma that enters the renal circulation.

In summary, clearance of a substance can range from zero to renal plasma flow, depending on how the substance is handled by the kidney.

Clearance = GFR	Filtration Only
Clearance < GFR	Filtration + Reabsorption
Clearance > GFR	Filtration + Secretion

CONTROL OF BLADDER AND LOWER URINARY TRACT

Urine produced by the kidneys is stored in the urinary bladder and periodically released through the urethra during urination. The three muscular structures of lower urinary tract—urinary bladder, bladder outlet, and external urethral sphincter—are regulated by the three branches of the nervous system. Smooth muscle cells in the urinary bladder and internal urethral sphincter in the bladder outlet are regulated by the pelvic nerve of the parasympathetic nervous system and the hypogastric nerve of the sympathetic nervous system. Detrusor smooth muscle cells in the urinary bladder contract in response to parasympathetic stimulation. Smooth muscle cells in the internal urethral sphincter contract in response to sympathetic stimulation. Skeletal muscle cells in the external urethral sphincter contract in response to somatic stimulation. As shown in Fig. 8.12, skeletal muscle cells in the external urethral sphincter are regulated by the pudendal nerve of the somatic nervous system.

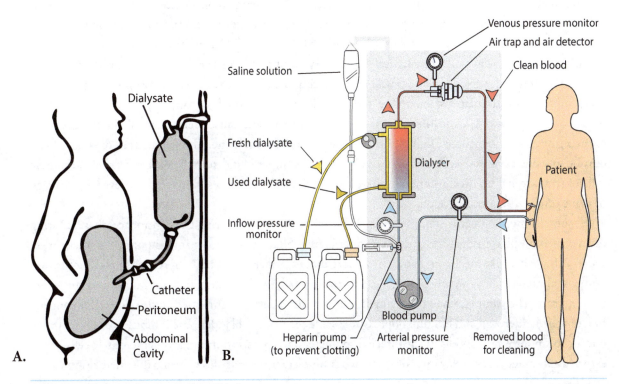

Fig. 8.13—A. Peritoneal dialysis.   **B.** Hemodialysis.

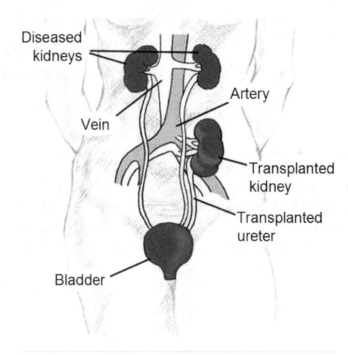

Fig. 8.14—Kidney transplantation.

The micturition center in the brain stem regulates urine storage (continence) and urination (micturition) of the lower urinary tract by regulating parasympathetic, sympathetic, and somatic outputs in response to mechanosensory input from the urinary bladder wall. During the continence phase, when the urinary bladder is not fully filled, mechanosensory input to the micturition center is low, because mechanical stretch of the urinary bladder wall is low. In response, the micturition center inhibits parasympathetic output and stimulates sympathetic and somatic outputs to the lower urinary tract. The inhibition of parasympathetic output to the detrusor smooth muscle cells causes relaxation of the bladder wall for accommodating continued filling of the urinary bladder with urine. The stimulation of sympathetic output to smooth muscle cells of the internal urethral sphincter and somatic output to skeletal muscle cells of the external urethral sphincter causes contraction of these two sphincters for preventing leakage of urine during the continence phase.

Events during the micturition phase are opposite to the events during the continence phase. During the micturition phase, when the urinary bladder is fully filled, mechanosensory input to the micturition center is high, because mechanical stretch of the urinary bladder wall is high. In response, the micturition center stimulates parasympathetic output and inhibits sympathetic and somatic outputs to the lower urinary tract. The stimulation of parasympathetic output to detrusor smooth muscle cells causes contraction of the bladder wall and buildup of pressure inside the urinary bladder. The inhibition of sympathetic output to smooth muscle cells of the internal sphincter and somatic output to skeletal muscle cells of the external sphincter causes relaxation of these two sphincters for urinary flow during the micturition phase.

Overactive bladder syndrome is characterized by an increase in urgency for urination. For treatment of overactive bladder, muscarinic receptor antagonists are commonly used for blocking the stimulating effect of parasympathetic nervous system on detrusor smooth muscle cells in the bladder wall. New drugs targeting the sensory pathway from urinary bladder to the micturition center are being developed for the treatment of overactive bladder syndrome.

KIDNEY FAILURE AND DIALYSIS

Chronic kidney failure is characterized by a diminished glomerular filtration rate, increased urinary excretion of protein, or both. Kidney failure leads to the accumulation of waste products and excess volume in the extracellular fluid, which can lead to failures in multiple organ systems. For example, an abnormally high concentration of K^+ can lead to abnormalities in membrane potentials and cardiac arrhythmia.

Dialysis systems are used for removing waste products from the extracellular fluid in patients having kidney failure. The basic principle of dialysis is a diffusional exchange of waste products between extracellular fluid and dialysate through a membrane system that is selectively permeable to small molecules but impermeable to proteins and cells. During peritoneal dialysis, as shown in Fig. 8.13A, clean dialysate is infused into the abdominal cavity via a catheter. The peritoneum lining the abdominal cavity consists of a monolayer of mesothelial cells facing the abdominal cavity and a capillary network underneath the mesothelial layer. The peritoneum serves as a membrane system for diffusional exchange between the dialysate and the extracellular fluid. After four to six hours of equilibration, the dialysate is drained from the abdominal cavity into a drain bag for disposal. Peritoneal dialysis can be performed at home. A typical schedule for peritoneal dialysis is four times each day.

During hemodialysis, as shown in Fig. 8.13B, blood is pumped out of the patient's circulation via an artery into a membrane dialyzer for diffusional exchange with dialysate. Cleansed blood leaving the membrane dialyzer is pumped back into the patient's circulation via a vein. The dialyzer membrane is engineered to be permeable to small molecules but impermeable to proteins and cells. Dialysate is continuously refreshed with clean dialysate by a separate pump. Hemodialysis is usually performed in a clinical center. Each hemodialysis session lasts from three to five hours. A typical schedule for hemodialysis is three times a week.

Kidney transplantation is a surgical procedure for transplanting a healthy kidney from a donor to the recipient, as shown in Fig. 8.14. The donor kidney's renal artery and renal vein are typically connected to the recipient's artery and vein in the abdominal cavity. The donor kidney's ureter is typically connected to the recipient's urinary bladder. The recipient's failed kidneys are usually left in place, unless they cause infection or hypertension in the patient.

KEY TERMS

- acid–base regulation
- afferent arteriole
- antidiuretic hormone
- countercurrent multiplication
- creatinine
- distal tubule
- efferent arteriole
- glomerular filtration
- hemodialysis
- juxtaglomerular apparatus
- kidney
- kidney failure
- kidney transplantation
- Loop of Henle
- macula densa
- nephron
- peritoneal dialysis
- proximal tubule
- renal blood flow
- renal clearance
- renal reabsorption
- renal secretion
- renin-angiotensin-aldosterone system
- vasopressin

IMAGE CREDITS

9

GASTROINTESTINAL PHYSIOLOGY AND REGULATION OF SUBSTRATE METABOLISM

Normal function of the gastrointestinal tract is essential for growth and survival. For example, children with inflammatory bowel disease exhibit retarded growth and delayed puberty. Conversely, excessive food intake leads to obesity and metabolic diseases such as type 2 diabetes mellitus. This chapter addresses functions of the gastrointestinal tract, control of appetite, and regulation of nutrient metabolism.

STRUCTURE OF THE GASTROINTESTINAL TRACT

As shown in Fig. 9.1, the gastrointestinal tract is essentially a tube connecting the mouth to the anus. Major segments of the gastrointestinal tract include the mouth, esophagus, stomach, small intestine, large intestine, rectum, and anus.

Sphincters Along the Gastrointestinal Tract. The following sphincters regulate the movement of content between consecutive segments along the gastrointestinal tract.

LEARNING OBJECTIVES

1. **Gastrointestinal Motility.** Describe the neural and local mechanisms of swallowing, peristalsis, segmentation, and defecation.

2. **Gastrointestinal Secretions.** Describe the secretory function of chief cells and parietal cells in the stomach and their regulation by autonomic nervous system, gastrointestinal endocrine cells, and local cells.

3. **Pancreatic and Gall Bladder Secretions.** Describe the major products of pancreatic secretion and their regulation by negative feedback; describe the regulation of bile release from gall bladder to duodenum by negative feedback.

4. **Carbohydrate, Protein, and Fat Digestion and Absorption.** Describe the mechanisms of digestion and absorption of carbohydrates, protein, and fat.

5. **Regulation of Substrate Metabolism.** Describe the major regulatory hormones and typical measurements of plasma glucose homeostasis; compare and contrast the pathophysiology of type 1 and type 2 diabetes mellitus.

6. **Regulation of Satiety.** Define and provide examples of appetite suppressors and appetite stimulants for the regulation of food intake.

Mouth → Esophagus:	Upper Esophageal Sphincter
Esophagus → Stomach:	Lower Esophageal Sphincter
Stomach → Duodenum:	Pyloric Sphincter
Ileum → Cecum:	Ileocecal Valve
Rectum → Exterior:	External and Internal Anal Sphincters

Digestive Organs. The liver, gall gladder, and pancreas are digestive organs attached to the gastrointestinal tract for producing **bile**, digestive enzymes, and bicarbonate for digestion, as shown in Fig. 9.1. The liver synthesizes bile and releases bile to the gall bladder for storage. During digestion,

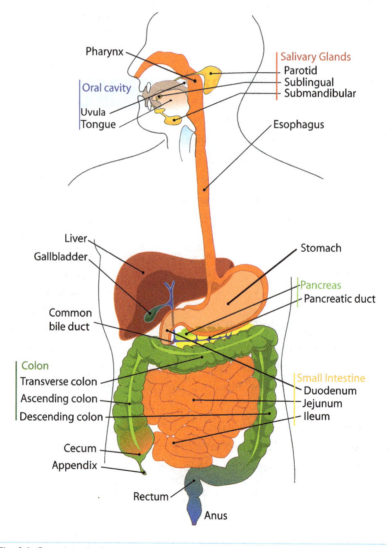

Fig. 9.1. Gastrointestinal system.

the gall bladder releases bile into the duodenum for fat digestion. Bile functions as detergent for breaking up fat into small micelles for digestion. The pancreas secretes digestive enzymes and bicarbonate into the duodenum of the small intestine for digestion of carbohydrates, proteins, and fat.

Gastrointestinal and Hepatic Circulation. Mesenteric arteries carry oxygenated blood from the abdominal aorta to the gastrointestinal tract, and the hepatic portal vein carries partially deoxygenated blood from the gastrointestinal tract to the liver, as illustrated in the following scheme:

Mesenteric Arteries → Gastrointestinal Tract → Hepatic Portal Vein → Liver

This anatomical arrangement implies that all substances absorbed by the gastrointestinal tract must first pass through the liver (first pass) before entering the general circulation. Liver is a major organ for drug metabolism. For drugs that are delivered orally, the first-pass effect of liver can significantly reduce the availability of a drug to the general circulation.

In addition, as shown in Fig. 9.2, the hepatic artery carries oxygenated blood from the abdominal aorta to the liver, and the central vein carries blood from the liver to the inferior vena cava. The following scheme summarizes the liver circulation:

Hepatic Portal Vein and Hepatic Artery → Liver → Hepatic Central Vein

As shown in Fig. 9.2, within the liver, hepatic arterial and portal venous blood are mixed in sinusoids lined by hepatocytes. Anatomically, the hepatic artery, portal vein, and bile duct are arranged together at the same location in the liver known as the portal triad.

GASTROINTESTINAL MOTILITY

Movement of food and digestive products through the gastrointestinal tract is regulated by multiple reflexes—swallowing reflex, **peristalic** reflex, and the defecation reflex.

Swallowing Reflex. As shown in Fig. 9.3, the trachea and esophagus share the oral-nasal cavity for the

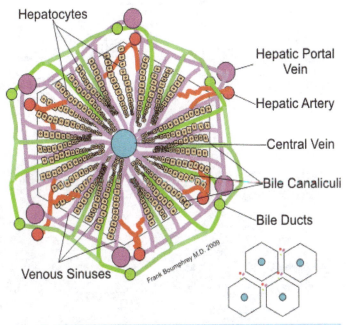

Fig. 9.2—Basic Structure of Liver Lobule. Liver receives blood from the hepatic artery, a branch of the abdominal aorta, and portal vein, which drains blood from the intestines. The liver produces bile, which is stored in the gall bladder. Portal triad consists of portal vein, hepatic artery, and bile duct. Blood from portal vein and hepatic artery enters sinusoids lined by hypatocytes, and leaves the sinusoids via the central vein, which drains into the inferior vena cava.

passage of air and food. Under the control of the swallowing center in the brain stem, the swallowing reflex regulates closing of the entry to the trachea and opening of the upper esophageal sphincter for the entry of food from mouth into the esophagus. Swallowing is initiated by voluntary movement of food bolus toward the back of the oral cavity, where the food bolus triggers sensory input into the swallowing center. At the same time, higher centers of the cerebral cortex also send input to the swallowing center. In response, the swallowing center executes a pattern generator program that sends inhibitory signals to the respiratory center to inhibit respiration, excitatory signals to muscle groups in the larynx-pharynx region to cause downward tilting of the epiglottis to cover the airways, and inhibitory signals to the upper esophageal sphincter to induce relaxation of the sphincter, thereby allowing the entry of food bolus into the esophagus.

The Peristaltic Reflex propels food and digestive products along the gastrointestinal tract from oral to anal direction. As shown in Fig. 9.4A, **peristalsis** involves contraction of the gastrointestinal tract behind the bolus and relaxation of the gastrointestinal tract in front of the bolus. As shown in Fig. 9.4B, local enteric neurons, under the modulation of the central and enteric nervous systems, regulate the coordinated contraction and relaxation of the gastrointestinal tract during peristalsis.

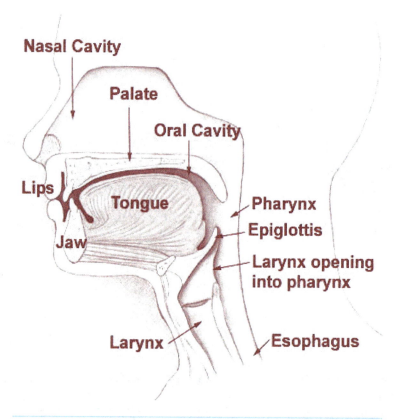

Fig. 9.3—Structure of the Oral Cavity and its Connection to the Trachea via Larynx and Esophagus via Pharynx. During swallowing, the epiglottis covers the opening of the larynx to prevent food from entering the trachea.

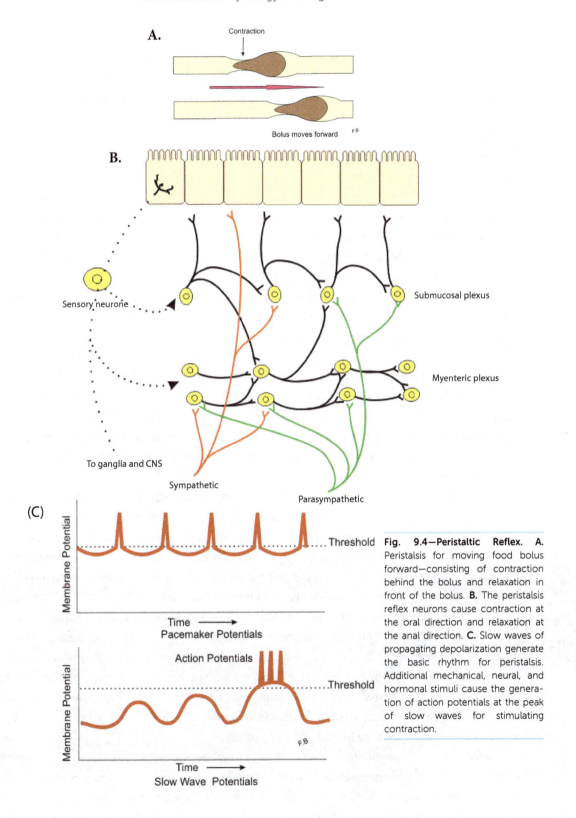

A. Contraction

Bolus moves forward

B.

Sensory neurone

Submucosal plexus

Myenteric plexus

To ganglia and CNS

Sympathetic

Parasympathetic

(C)

Membrane Potential

Threshold

Time →
Pacemaker Potentials

Action Potentials

Threshold

Membrane Potential

Time →
Slow Wave Potentials

Fig. 9.4—Peristaltic Reflex. A. Peristalsis for moving food bolus forward—consisting of contraction behind the bolus and relaxation in front of the bolus. **B.** The peristalsis reflex neurons cause contraction at the oral direction and relaxation at the anal direction. **C.** Slow waves of propagating depolarization generate the basic rhythm for peristalsis. Additional mechanical, neural, and hormonal stimuli cause the generation of action potentials at the peak of slow waves for stimulating contraction.

Peristalsis in the human esophagus is mediated by both skeletal and smooth muscle contractions, because the upper third of the esophagus contains skeletal muscle cells, whereas the lower two-thirds of the esophagus contains smooth muscle cells. Peristalsis in the skeletal muscle portion of the esophagus is regulated by coordinated sequential activation of motor neurons by the swallowing center. Peristalsis in the smooth muscle portion of the esophagus consists of voluntarily initiated (primary) and involuntarily initiated (secondary) peristalsis. Primary peristalsis in the smooth muscle portion of the esophagus is regulated by sequential activation of smooth muscle cells by the parasympathetic nervous system. Secondary peristalsis in the smooth muscle portion of the esophagus and other parts of the gastrointestinal tract is regulated by the enteric nervous system under modulation by hormones and the autonomic nervous system. During secondary peristalsis, pacemaker cells in the gastrointestinal tract (interstitial cells of Cajal) generate the slow waves of depolarization that propagate along the gastrointestinal tract, as shown in Fig. 9.4C. The slow waves of depolarization set the basic rhythm of peristalsis by periodically bringing the membrane potential of smooth muscle cells close to the threshold for generating action potentials at the peaks of slow waves. For triggering peristalsis, mechanical stretch and excitatory neurotransmitters are necessary for causing depolarization of smooth muscle cells to be above the threshold for triggering action potentials. For example, parasympathetic stimulation is excitatory for gastrointestinal motility. In comparison, sympathetic stimulation is inhibitory for gastrointestinal motility.

Gastric Accommodation and Gastric Emptying. Gastric accommodation (also known as receptive relaxation) refers to the relaxation of the stomach in response to the entry of food during swallowing. Gastric accommodation allows the stomach to store a large amount of food with minimal increase in transmural pressure. Gastric accommodation is mediated by the enteric and parasympathetic nervous systems.

Gastric peristalsis performs two functions: mixing of food for digestion, and emptying of gastric content into the duodenum. Gastric emptying via the pyloric sphincter is determined by the physical and chemical characteristics of gastric content—physical state, caloric content, tonicity, and temperature. For example, gastric emptying of liquid is faster than that of solids. When both liquid and solids are present, gastric emptying of liquid is completed first. Gastric emptying of solids is typically delayed until the large solids are broken into small particles. Gastric emptying is regulated by caloric content at approximately one to four kilocalories per minute for optimal absorption by the small intestine. Gastric emptying is temperature-dependent. A decrease in temperature slows gastric emptying, and vice versa. Gastric emptying of fat is slower than gastric emptying of carbohydrate and protein, because the caloric content of fat is highest among these three substrates. Gastric emptying of isotonic content is faster than gastric emptying of hypertonic and hypotonic contents. Gastric emptying is inhibited by an intestinal hormone, **cholecystokinin (CCK)**, which is released by the small intestine in response to luminal fat and protein.

Defecation—expulsion of rectal content to the exterior through the anus—is important for the elimination of undigested products from the gastrointestinal tract. Normal frequency of defecation ranges widely from three times a week to three times a day. Defecation is regulated by both autonomic and somatic nervous system. As shown in Fig. 9.5, the parasympathetic nervous system regulates the

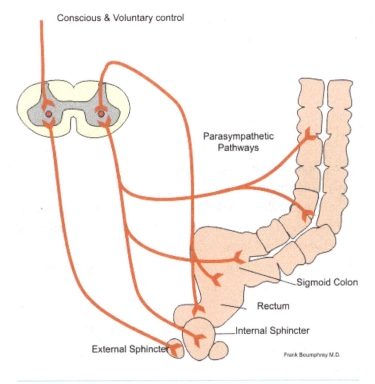

Conscious & Voluntary control

Parasympathetic
Pathways

Sigmoid Colon

Rectum

Internal Sphincter

External Sphincter

Frank Boumphrey M.D.

Fig. 9.5. Defecation reflex.

contraction of colon, rectum, and internal anal sphincter, and the somatic nervous system regulates the contraction of the external anal sphincter for voluntary control of defecation. The internal anal sphincter is also innervated by the sympathetic nervous system. The parasympathetic nervous system predominantly inhibits the contraction of the internal anal sphincter, whereas the sympathetic nervous system predominantly stimulates contraction of the internal anal sphincter.

Defecation consists of four phases: basal phase, pre-defecation phase, expulsion phase, and termination phase, the anal canal is closed during the basal phase due to tonic contractions of the internal and external anal sphincters. The internal anal sphincter is primarily responsible for resting anal continence. The skeletal muscle cells forming the external anal sphincter are unique in having a resting tone at rest, because most skeletal muscle cells are in the relaxed state at rest. Colonic peristalsis propels gastrointestinal content from the colon into the rectum, causing accumulation of content in the rectum and rectal distension. During the pre-defecation phase, rectal distention activates the mechanosensitive sensory pathway to the central nervous system, where it triggers the recto-anal inhibitory reflex, resulting in the urge for defecation, relaxation of the internal anal sphincter, and contraction of the external anal sphincter for maintaining continence. Defecation can be delayed as long as anal pressure exceeds rectal pressure. During the expulsion phase, rectal pressure increases as a result of voluntary straining of abdominal muscles and involuntary colorectal contraction. At the same time, anal pressure decreases as a result of voluntary relaxation of the external anal sphincter and pelvic floor muscle, and involuntary relaxation of the internal anal sphincter as a result of the recto-anal inhibitory reflex. Expulsion of rectal content to the exterior occurs when rectal pressure exceeds anal pressure. During the termination phase, the sense of complete rectal evacuation causes the cessation of activities for defecation and return of the internal and external anal sphincters to the state of contraction.

SECRETORY FUNCTION OF THE GASTROINTESTINAL TRACT

Cells situated at different segments of the gastrointestinal tract secrete mucous, digestive enzymes, hydrochloric acid (HCl), bicarbonate, and hormones. All segments of the gastrointestinal tract secrete mucous for protecting epithelial cells from digestive enzymes and reducing friction for the movement of content along the gastrointestinal tract. The stomach specializes in secreting HCl and pepsinogen for protein digestion and intrinsic factor for intestinal absorption of vitamin B_{12}. The pancreas specializes in secreting bicarbonate and enzymes into the small intestine for neutralization of acid and digestion of carbohydrates, proteins, and fat. The liver secretes bicarbonate and bile into the small intestine for neutralization of acid and emulsification of fat. Secretions by different segments of the gastrointestinal are examined in the following paragraphs.

Oral Secretions. Salivary glands secrete amylase into the oral cavity for the digestion of polysaccharides—for example, starch, into mono- and di-saccharides. In principle, digestion of polysaccharides begin in the mouth, but the extent of starch digestion in the oral cavity is limited, because food stays in the mouth for a relatively short period before being swallowed into the esophagus.

Gastric Secretions. Fig. 9.6 shows the major secretory cells in gastric pits in the stomach. **G cells** are endocrine cells that secrete the hormone gastrin into the circulation for the regulation of gastric secretions. **Chief cells** secrete the proenzyme pepsinogen into the stomach, where pepsinogen is converted by acid to the active enzyme pepsin for protein digestion. Pepsin also catalyzes the conversion of pepsinogen to pepsin, thereby amplifying the effect of acid on pepsin formation. **Parietal cells** secrete HCl for the activation of pepsinogen and **intrinsic factor** (a 60,000-dalton carrier protein) for intestinal absorption of vitamin B_{12}. Vitamin B_{12} is a 1,400-dalton cobalt-containing, water-soluble, membrane-impermeable molecule,

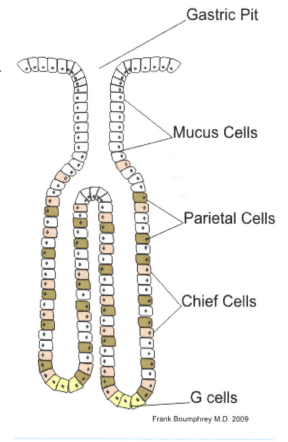

Gastric Pit

Mucus Cells

Parietal Cells

Chief Cells

G cells

Frank Boumphrey M.D. 2009

Fig. 9.6. Secretory cells in gastric pit.

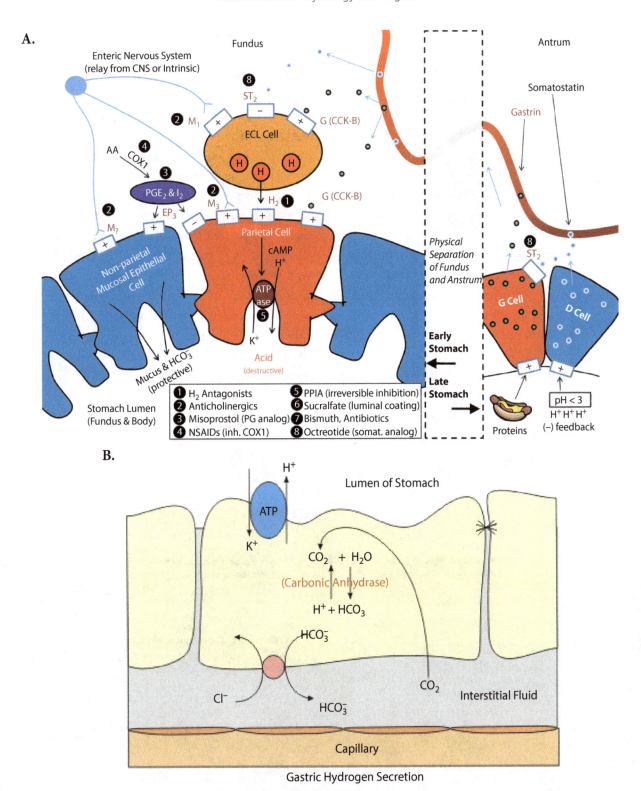

Fig. 9.7—A. Regulation of acid secretion in the stomach.   **B.** Transporters for acid secretion by parietal cells.

essential for DNA synthesis and cellular energy production. The intrinsic factor-vitamin B_{12} complex formed in the stomach is transported to the small intestine, where the intrinsic factor-vitamin B_{12} complex is endocytosed by intestinal epithelial cells for lysosomal degradation and transport into the circulation.

Phases of Gastric Acid Secretion. Gastric acid secretion consists of three phases: cephalic phase, gastric phase, and intestinal phase. The cephalic phase refers to the stimulation of gastric acid secretion by the anticipation of food intake. The cephalic phase of gastric acid secretion is mediated by parasympathetic stimulation of parietal cells. The post-ganglionic parasympathetic neurotransmitter, acetylcholine, activates muscarinic acetylcholine receptors on parietal cells.

The gastric phase refers to the stimulation of gastric acid secretion by the presence of food in the stomach. The gastric phase of gastric acid secretion is regulated by gastric hormones, a local mediator, and the enteric nervous system. As shown in Fig. 9.7A, G cells in the stomach secrete the gastric hormone gastrin into the circulation. **Gastrin** directly stimulates gastric acid production by activating gastrin receptors on parietal cells. In addition, gastrin indirectly stimulates gastric acid production by stimulating the release of histamine by **enterochromaffin-like cells** (ECL) to the interstitium, where histamine stimulates acid production by parietal cells by activating histamine H_2 receptors. **D cells** in the stomach (Fig. 9.7A) function as a negative feedback mechanism by releasing the inhibitory gastric hormone **somatostatin** in response to acidity in the stomach. Somatostatin

A.

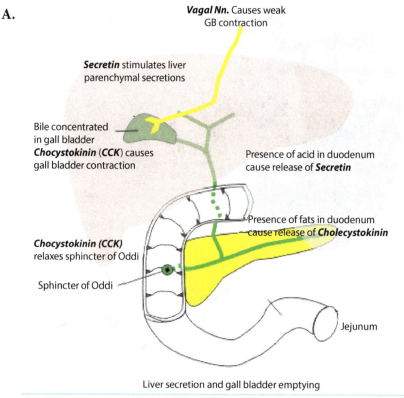

Liver secretion and gall bladder emptying

Fig. 9.8—A. Control of pancreatic secretions.   B. Secretion and reabsorption of bile salts.

inhibits acid secretion by parietal cells by binding to somatostatin receptors. Enteric neurons release acetylcholine that activates muscarinic acetylcholine receptors on parietal cells.

The intestinal phase of gastric acid secretion refers to the inhibition of gastric acid production in response to the presence of nutrients in the duodenum, the first section of the small intestine. Intestinal phase of gastric acid secretion is mediated by the enteric nervous system and intestinal hormones (CCK, secretin, and others) in response to the presence of nutrients in the duodenum.

Molecular Mechanism of HCl Secretion by Parietal Cells. Fig. 9.7B shows the molecular mechanism of hydrochloric acid production by a parietal cell. Intracellular carbonic anhydrase catalyzes the reaction of carbon dioxide with water to form carbonic acid, which dissociates into H^+ and HCO_3^-. H^+ is pumped into the stomach lumen by the active transporter H^+-K^+-ATPase on the luminal membrane of parietal cell. K^+ is also pumped into the cell, but diffuses back to the stomach lumen via K^+ channels on the luminal membrane. HCO_3^- is transported out of the cell into the circulation in exchange with Cl^- transported from the circulation into the cell by the **HCO_3^-/ Cl^- exchanger** on the basolateral membrane of the parietal cell. Cl^- then diffuses from the cell into the stomach lumen via chloride channels. The net result is the production of HCl into the stomach lumen. During gastric acid production, the transport of HCO_3^- into the circulation causes alkalinization of blood leaving the stomach, which is known as "alkaline tide."

B.

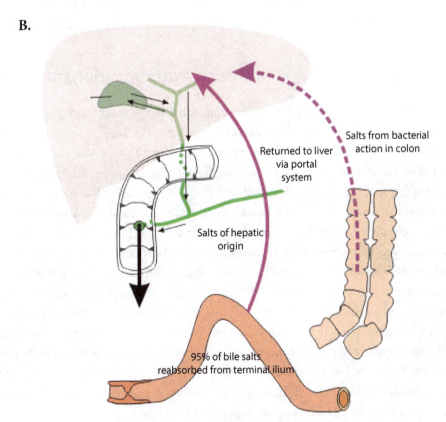

Returned to liver via portal system

Salts from bacterial action in colon

Salts of hepatic origin

95% of bile salts reabsorbed from terminal ilium

Enterophepatic circulation of bile salts

Pancreatic Secretions. The pancreas is the most important source of digestive enzymes—amylase, protease, and lipase—for the digestion of carbohydrate, protein, and fat. The pancreas secretes bicarbonate and enzymes into the duodenum, the first segment of the small intestine, for neutralization of acid and digestion of carbohydrates, proteins, and fat. Two negative feedback loops involving two intestinal hormones—secretin and cholecystokinin (CCK)—regulate pancreatic secretions of bicarbonate and enzymes in response to intestinal contents. In the first negative feedback loop, as shown in Fig. 9.8A, high acidity in the duodenum induces intestinal secretion of the hormone secretin into the circulation. Secretin stimulates pancreatic secretion of bicarbonate for the neutralization of acid, thereby decreasing the acidity in the duodenum. In the second negative feedback loop, the presence of fat in the duodenum induces intestinal secretion of the hormone CCK into the circulation. CCK stimulates pancreatic secretion of enzymes for the digestion of carbohydrates, proteins, and fat, thereby decreasing fat content in the duodenum.

Release of Bile by the Gall Bladder. Bile salts function as a detergent for breaking up fat into small micelles, thereby increasing the surface area for digestion by pancreatic lipases. As shown in Fig. 9.8B, bile salts are synthesized in the liver and transported to the gall bladder for storage. The presence of fat in the duodenum induces intestinal secretion of the hormone CCK into the circulation. CCK stimulates the contraction of the gall bladder and relaxation of the sphincter of Oddi for the release of bile into the duodenum for fat digestion (Fig. 9.8A). After intestinal absorption of fat, 95 percent of bile salts are reabsorbed by the ileum, the last segment of the small intestine, and returned to the liver via the portal vein (Fig. 9.8B). Approximately 5 percent of bile salts are excreted.

DIGESTIVE AND ABSORPTIVE FUNCTIONS OF THE GASTROINTESTINAL TRACT

The small intestine is the most important gastrointestinal segment for the digestion of carbohydrate, protein, and fat, and absorption of the digested products. The small intestine has a huge amount of surface area for digestion and absorption due to the presence of villi (macroscopic finger-like projections) on the intestinal surface and microvilli (microscopic finger-like projections) on the cell membrane of intestinal epithelial cells. As shown in Fig. 9.9A, each villus contains intestinal epithelial cells on the surface of the villus for digestion and absorption, and blood and lymphatic capillaries in the middle of villus for the absorption of digested products. Epithelial cells on intestinal villi are constantly renewed by the migration of new cells that are produced by stem cells at the bottom of intestinal crypts. The cell renewal process involves cell proliferation at the bottom of the crypt, cell differentiation near the top of the crypt, cell migration from the crypt to the villus, and cell extrusion at the tip of the villus. Globlet cells secrete mucus that lubricates the intestinal tract for peristaltic movement of content along the intestine. As shown in Fig. 9.9B, intestinal epithelial cells contain microvilli—finger-like projections from the cell membrane—facing the intestinal lumen. Microvilli, also known as brush border, substantially increase the surface area for the absorption of nutrients by intestinal epithelial cells. Microvilli also contain membrane-anchored enzymes for the digestion of disaccharides to monosaccharides and small peptides to amino acids.

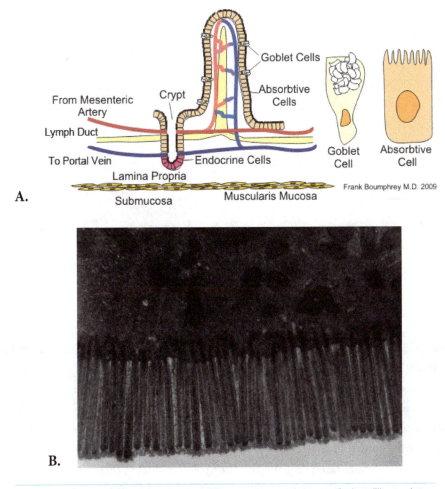

Fig. 9.9—A. Structure of an intestinal villi.   **B.** Electron microscopic image of microvilli on an intestinal cell.

Carbohydrate Digestion and Absorption. Fig. 9.10 summarizes the mechanisms of carbohydrate digestion and absorption in the small intestine. Digestion of complex carbohydrates consists of two separate enzymatic reactions—breakdown of polysaccharide to disaccharide and breakdown of disaccharide to monosaccharide. Salivary and pancreatic amylases catalyze the breakdown of polysaccharide (starch and glycogen) into disaccharides (lactose, maltose, and sucrose). Disaccharidases, situated on the brush borders of intestinal epithelial cells, catalyze the breakdown of disaccharides into monosaccharides. As shown in Fig. 9.10, the brush-border enzyme lactase catalyzes the breakdown of the milk sugar, lactose, into glucose and galactose. Most adults in the world population are lactase-deficient. Consumption of untreated milk by lactase-deficient adults can lead to the passage of lactose to the large intestine, where bacterial consumption of lactose can lead to flatulence and diarrhea. The brush-border enzyme maltase catalyzes the breakdown of maltose into glucose. The brush-border enzyme sucrase catalyzes the breakdown of sucrose (common sugar) into fructose and glucose.

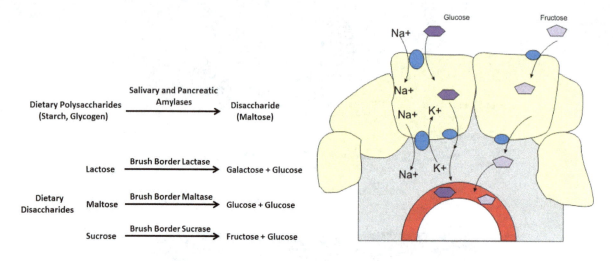

Dietary Polysaccharides   Salivary and Pancreatic
(Starch, Glycogen)   Amylases   Disaccharide
                                               (Maltose)

Lactose   Brush Border Lactase   Galactose + Glucose

Dietary
Disaccharides   Maltose   Brush Border Maltase   Glucose + Glucose

Sucrose   Brush Border Sucrase   Fructose + Glucose

Fig. 9.10. Intestinal carbohydrate digestion and absorption.

As shown in Fig. 9.10, glucose and galactose are transported into intestinal epithelial cells by the SGLT1 sodium-glucose cotransporter, situated on the luminal membrane of the cell. This secondary active transport, driven by the Na^+ gradient, enables complete intestinal absorption of galactose and glucose against their concentration gradients and establishing high concentrations of galactose and glucose inside intestinal epithelial cells. Glucose and galactose inside intestinal epithelial cells diffuse down their concentration gradients via the GLUT2 glucose carrier, found on the basolateral membrane of intestinal epithelial cells, to enter the circulation. In contrast, fructose enters intestinal epithelial cells passively by facilitated diffusion via the GLUT5 glucose carrier on the luminal membrane and then diffuses down its concentration gradient via the GLUT2 glucose carrier, situated on the basolateral membrane of intestinal epithelial cells. In addition, some fructose is converted to glucose by isomerase inside intestinal epithelial cells. The passive mechanism for intestinal absorption of fructose limits human tolerance of fructose intake. Excessive intake of fructose can result in an accumulation of residual fructose in the gastrointestinal tract for bacterial fermentation, flatulence, and abdominal discomfort.

Protein Digestion and Absorption. As shown in Fig. 9.11, exogenous proteins from food and endogenous proteins from dead cells and enzymes are digested by enzymes released by exocrine glands in the stomach and pancreas and enzymes found on the brush border of intestinal epithelial cells. Protein digestion begins in the stomach and completes in the small intestine. As cited previously, chief cells in the stomach secrete the proenzyme pepsinogen, which is converted to the active enzyme pepsin by the high acidity in the stomach. The acidic environment is also optimal for the function of pepsin. Protein digestion by pepsin and mechanical churning by the stomach convert ingested food into chyme, a mixture of fluid and fine particles of food. Opening of the pyloric sphincter releases chyme into the duodenum, where acid in chyme is neutralized by bicarbonate secreted by the pancreas, thereby optimizing the pH for digestion by pancreatic enzymes. As shown in Fig. 9.11, pancreatic proteases and brush-border protease together catalyze the breakdown of proteins into dipeptides, small peptides, and amino acids. Dipeptides and small peptides are transported from the lumen into intestinal epithelial cells via peptide transporters—for example, PEPT1, found on the luminal membrane, and then broken down by intracellular peptidases into amino acids. Amino acids are transported from the lumen into intestinal epithelial cells via Na⁺-amino acid cotransport and carrier-mediated facilitated diffusion. Amino acids are transported from intestinal epithelial cells to the circulation via carrier-mediated facilitated diffusion. Some large peptides and proteins are transported across intestinal epithelial cells by endocytosis at the luminal membrane and exocytosis at the basolateral membrane.

Fat Digestion and Absorption. Digestion of fat is more complicated than the digestion of carbohydrates and proteins, because fat, being water-insoluble, requires bile for dispersion into micelles. Bile salts are synthesized in the liver, stored in the gall bladder, and released into the duodenum. Bile functions as a detergent in dispersing fat into small micelles (emulsion droplets), thereby increasing the surface area for digestion. Fat is broken down into free fatty acids and monoglycerides by pancreatic lipase. Small and medium-chain (4–12 carbon) fatty acids can diffuse across the luminal and basolateral membranes of intestinal epithelial cells into blood capillaries. In comparison, as shown in Fig. 9.12, long-chain free fatty acids and monoglycerides are transported into intestinal epithelial cells via both simple diffusion and fatty acid transport proteins (FATP), situated at the luminal membrane, and then resynthesized into triglycerides inside intestinal epithelial cells and assembled into chylomicrons for exocytosis. Cholesterol is similarly transported into intestinal cells by protein transporters at the luminal membrane and assembled into **chylomicrons** for exocytosis. Fig. 9.13

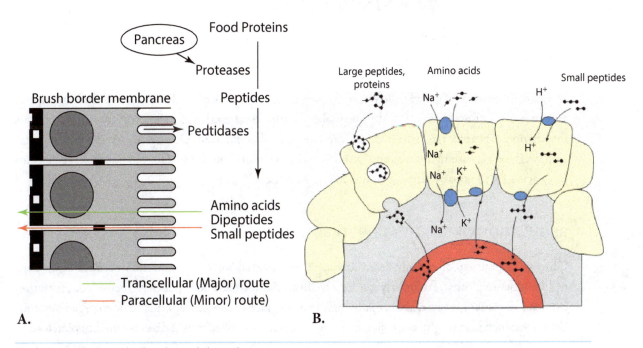

Fig. 9.11. Intestinal protein digestion and absorption.

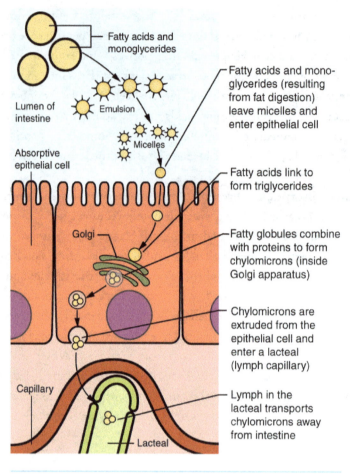

Fig. 9.12. Intestinal fat digestion and absorption.

shows the structure of a chylomicron, consisting of an inner core of triglycerides and cholesterol, and an outer shell of apoproteins and phospholipids. Intestinal epithelial cells exocytose chylomicrons into lymphatic capillaries, which drain into the general circulation. During transit through circulation, triglycerides are hydrolyzed by lipoprotein lipase, found on the surface of endothelial cells, into fatty acids for uptake by adipose tissue, liver, and muscle.

REGULATION OF SUBSTRATE METABOLISM

Glucose, a product of carbohydrate digestion, is essential for brain function, because brain cells utilize glucose almost exclusively for metabolism. Abnormally low plasma glucose concentration can lead to coma; however, abnormally high plasma glucose concentration is toxic to organ systems. Sustained high plasma glucose concentration in poorly controlled type 2 diabetes mellitus can lead to

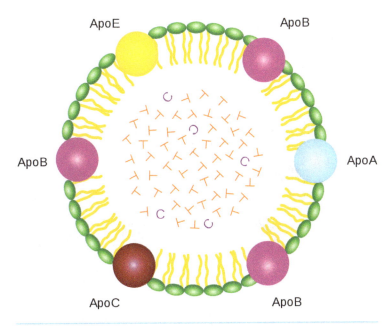

Fig. 9.13. Structure of a chylomicron, consisting of lipoproteins (ApoA, ApoB, ApoC, ApoE), phospholipids (green), triglyceride (T) and cholesterol (C).

kidney failure. Maintaining plasma glucose concentration within a normal range (4–8 mM; 72–144 mg/100 ml) is important for health.

The two pancreatic hormones, **glucagon** and **insulin**, are the two most important regulators of glucose homeostasis. As shown in Fig. 9.14, when plasma glucose concentration is abnormally low, pancreatic α cells secrete glucagon, which stimulates glycogenolysis—that is, a breakdown of glycogen to glucose, in the liver, thereby increasing plasma glucose concentration to normal level. Glucagon also stimulates gluconeogenesis—the production of glucose from substrates other than glycogen— in the liver and kidney, thereby increasing glucose concentration in the plasma. For example, amino

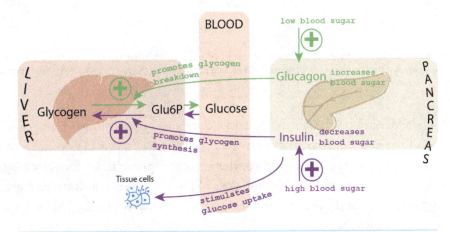

Fig. 9.14. Primary hormonal control of plasma glucose homeostasis.

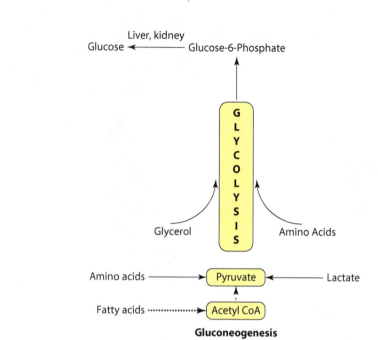

Gluconeogenesis

A.

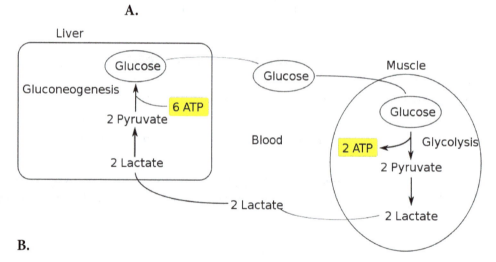

B.

Fig. 9.15—A. Gluconeogenesis for the production of glucose from amino acid, glycerol, lactate and pyruvate.   **B.** The Cori Cycle for conversion of lactate produced by muscle to glucose by gluconeogenesis in the liver.

acids, fatty acids, glycerol, and lactate can be used to synthesize glucose by gluconeogenesis, as shown in Fig. 9.15A. Liver and kidney are the only two organs that are capable of gluconeogenesis, because only these two organs contain the critical enzyme glucose-6-phosphatase for glucose synthesis. As shown in Fig. 9.15B, during heavy exercise, skeletal muscle cells break down glycogen to lactate, which is then circulated to the liver to produce glucose by gluconeogenesis. The shuttling of lactate from muscle to liver for the synthesis of glucose by gluconeogenesis is known as the Cori cycle. Sympathetic stimulation and other hormones—for example, epinephrine, cortisol, and growth

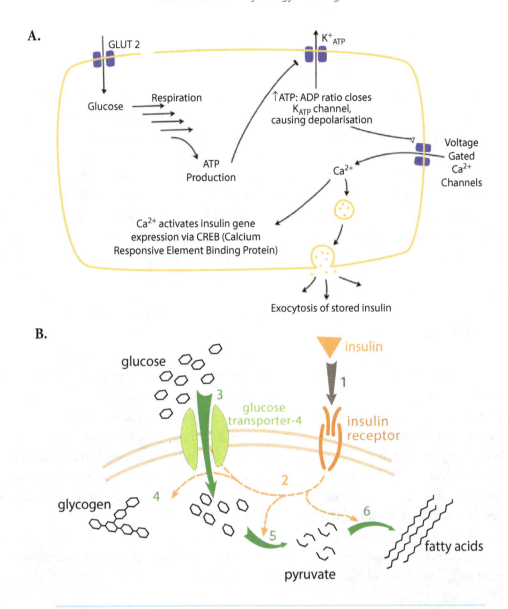

A.

GLUT 2

Glucose → Respiration

↑ATP: ADP ratio closes K_{ATP} channel, causing depolarisation

K^+_{ATP}

ATP Production

Voltage Gated Ca^{2+} Channels

Ca^{2+}

Ca^{2+} activates insulin gene expression via CREB (Calcium Responsive Element Binding Protein)

Exocytosis of stored insulin

B.

glucose

insulin

3

glucose transporter-4

1

insulin receptor

2

glycogen

4

5

6

pyruvate

fatty acids

Fig. 9.16—A. Mechanism of glucose-induced release of insulin by pancreatic β cells. **B.** Mechanism of insulin-induced glucose uptake by cells.

hormone—also stimulate the production of glucose by glycogenolysis and gluconeogenesis in the liver.

When plasma glucose concentration is high, for example, after eating a carbohydrate-rich meal, pancreatic β cells secrete insulin, which stimulates cellular uptake of glucose by liver and other organs for glycogen synthesis, thereby decreasing plasma glucose concentration to normal level. Insulin is the only hormone capable of decreasing plasma glucose concentration by stimulating glucose uptake by cells. As shown in Fig. 9.16A, the entry of glucose into pancreatic β cells via GLUT2 glucose carriers stimulates the secretion of insulin by pancreatic β cells by increasing intracellular ATP/

ADP ratio, which causes the closing of ATP-dependent K^+ channels, membrane depolarization, an opening of voltage-gated plasmalemmal Ca^{2+} channels, an increase in intracellular Ca^{2+} concentration, and Ca^{2+}-dependent exocytosis of insulin. As shown in Fig. 9.16B, binding of insulin to the insulin receptor on insulin-sensitive cells stimulates intracellular signaling pathways that lead to the expression of **GLUT4 glucose carriers** on the cell membrane for facilitated diffusion of glucose into cells for storage and metabolism, thereby lowering plasma glucose concentration.

Measurements of Glucose Homeostasis. Glucose at abnormally high plasma concentration is toxic to organ systems. For example, sustained hyperglycemia in type 2 diabetes mellitus is a major cause of kidney failure. Measurement of glucose homeostasis is important for the management of type 2 diabetes mellitus. The three common measures of glucose homeostasis are fasting plasma glucose concentration, blood level of glycated hemoglobin (Hb_{A1C}), and oral glucose tolerance test. Fasting plasma glucose concentration, typically measured after overnight fasting, is largely a measure of hepatic production of glucose. Fasting plasma glucose concentration higher than 126 mg/100 ml suggests diabetes mellitus. Fasting plasma glucose concentration is relatively easy to measure, but a single measurement of plasma glucose concentration at one time point is often insufficient for evaluating overall glucose homeostasis in weeks prior to the time of blood collection.

Glycated hemoglobin (Hb_{A1C}) is glucose-conjugated hemoglobin that is formed by a slow reaction between hemoglobin and glucose in the blood. The level of glycated hemoglobin in blood reflects relatively long-term glucose homeostasis, because the average life span of red blood cells is approximately one hundred twenty days. Normal level of Hb_{A1C} in nondiabetic subjects is less than 6 percent. A major goal in the treatment of patients having type 2 diabetes mellitus is to control Hb_{A1C} level at or below 6.5 percent. Hb_{A1C} is an important measure of long-term glucose homeostasis, but does not directly address the body's control of plasma glucose concentration at the time of measurement.

Oral glucose tolerance test measures the time course of plasma glucose concentration after oral ingestion of a standard amount (150 g) of glucose. A plasma glucose concentration higher than 200 mg/100 ml at two hours after glucose intake is an indicator of impaired glucose tolerance. Oral glucose tolerance test is a sensitive test of the body's control of plasma glucose concentration and an important indicator of the risk for developing diabetes mellitus.

DIABETES MELLITUS

Diabetes Mellitus is a disease state characterized by an abnormally high glucose concentration in the plasma and urinary excretion of glucose. **Type 1 diabetes mellitus** is an autoimmune disease, in which pancreatic β cells are destroyed by the immune system. In type 1 diabetes mellitus, in the absence of insulin, plasma glucose concentration is abnormally high, but intracellular glucose concentration is abnormally low, because cells are unable to take up glucose in the absence of GLUT4 glucose carriers on the cell membrane. The abnormally low intracellular glucose concentration in adipose cells causes an excessive increase in the breakdown of triglyceride into free fatty acids, which are metabolized by the liver to ketone bodies (β-hydroxybutyrate and acetoacetone). High levels of

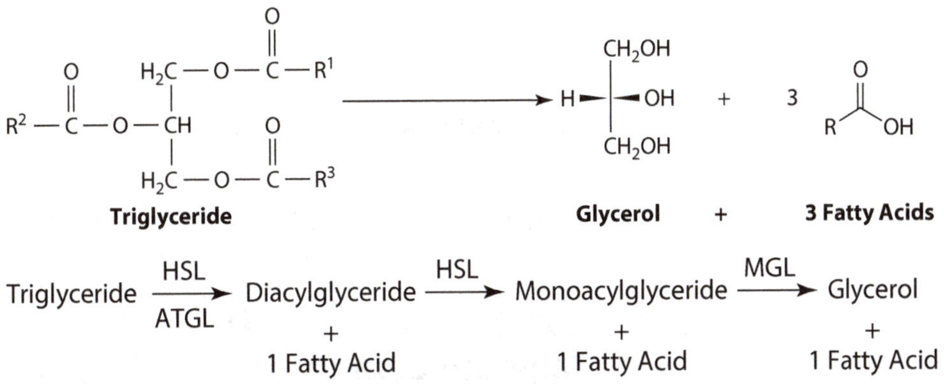

Fig. 9.17. Regulation of lipolysis in fat cells by hormone-sensitive lipase (HSL), adipose triglyceride lipase (ATGL), and monoglyceride lipase (MGL).

ketone bodies can lead to ketoacidosis and diabetic coma, and death. Patients having type 1 diabetes mellitus are typically treated with subcutaneous injections of insulin.

Type 2 Diabetes Mellitus is characterized by an insulin resistance-caused by dysfunction in the insulin receptor-coupled signaling pathways. Despite the presence of insulin, cells fail to insert sufficient GLUT4 glucose carriers onto the cell membrane for glucose uptake, causing plasma glucose concentration to become abnormally high. Obesity is a significant risk factor for the development of type 2 diabetes mellitus. Most patients having type 2 diabetes mellitus exhibit abnormally high plasma glucose concentration without ketoacidosis, because the level of insulin in plasma is often sufficient for suppressing excess lipolysis. Long-term overstimulation of pancreatic β cells by high plasma glucose concentration can eventually lead to pancreatic β cell failure and the loss of insulin secretion. Recent findings indicate that patients having type 2 diabetes mellitus may develop a sudden decrease in insulin production and ketoacidosis. Patients having type 2 diabetes mellitus are usually treated with drugs for increasing insulin sensitivity of cells or stimulating insulin secretion by pancreatic β cells. Recently, inhibitors of the Na^+-glucose cotransporter SGLT2 have been developed for the treatment of type 2 diabetes mellitus by inhibiting renal reabsorption of glucose, thereby increasing urinary excretion of glucose and lowering plasma glucose concentration in patients. With the progressive decline in pancreatic β cell function in patients having type 2 diabetes mellitus, administration of insulin eventually becomes one of the therapeutic options.

Regulation of Lipolysis. Lipolysis, the breakdown of triglyceride (also known as triacylglycerol, or TAG) to nonesterified fatty acids (NEFAs) and glycerol, in fat cells is an important mechanism for supporting metabolism of organ systems, because fatty acids are utilized by most nonneuronal cells, including cardiac and skeletal muscle cells, for energy metabolism. In addition, as shown previously in Fig. 9.15, the liver and kidneys can convert glycerol and fatty acids to glucose for utilization by neuronal cells.

Fig. 9.17A shows the overall metabolism of triglyceride to glycerol and three fatty acids. As shown in Fig. 9.17B, a **hormone-sensitive lipase (HSL)** in fat cells catalyzes the breakdown of triglyceride

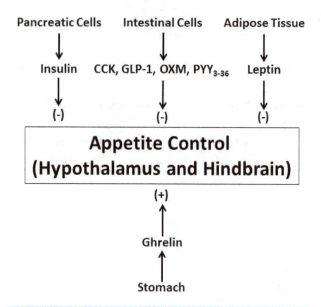

Fig. 9.18—Regulation of Satiety by Endogenous Hormones and Therapeutic Interventions. Abbreviations: CCK—cholecystokinin; GLP-1—glucagon-like peptide-1; OXM—oxyntomodulin; PYY_{3-36}—peptide YY residues 3–36.

to diacylglyceride and one fatty acid, and the breakdown of diacylglyceride to monoglyceride and one fatty acid. HSL in fat cells is activated by sympathetic stimulation, epinephrine, and glucagon, and inhibited by insulin. The cellular mechanism of HSL activation includes receptor-coupled activation of adenylate cyclase, generation of intracellular cyclic AMP (cAMP), activation of cAMP-dependent kinase, and HSL phosphorylation. During moderate exercise, fatty acid oxidation increases by five-fold to tenfold due to the availability of fatty acids produced by activated HSL in fat cells in response to high levels of sympathetic stimulation, epinephrine, and glucagon and low level of insulin. As shown in Fig. 9.17B, two other lipases—adipose triacylglycerol lipase (ATGL) and monoglycerol lipase (MGL)—also participate in intracellular lipolysis in fat cells. ATGL catalyzes the breakdown of triglyceride to diacylglyceride, whereas MGL catalyzes the breakdown of monoacylglyceride to glycerol and one fatty acid.

REGULATION OF SATIETY

Appetite for food intake is essential for survival. As shown in Fig. 9.18, multiple hormones and peptides released by gastrointestinal cells and adipocytes regulate the primary center for appetite control located in the hypothalamus. Most appetite-control molecules are appetite suppressors (anorexigenic molecules) that provide the satiety sensation after an enough amount of food has been eaten. Anorexigenic molecules include gastrointestinal hormones and peptides—for example, cholecystokinin (CCK), glucagon-like peptide-1 (GLP-1), oxyntomodulin (OXM), peptide YY (PYY), the pancreatic hormone insulin, and the adipose hormone leptin. Some appetite-control molecules are appetite stimulants (orexigenic molecules) that provide the hunger sensation during fasting. Orexigenic molecules include the gastric hormone ghrelin and the adipose hormone adiponectin. In general, eating food leads to a gradual increase in the concentration of anorexigenic molecules and a decrease in the concentration of orexigenic molecules, resulting in the sensation of satiety and suppression of further food intake. Fasting leads to a gradual decrease in the concentration of anorexigenic molecules and an increase in orexigenic molecules, resulting in the sensation of hunger and urge for food intake.

In addition to hormones and peptides, a glucose-sensing system in the hepatic portal vein detects glucose concentration in hepatic portal venous blood and sends signals via the vagus nerve to the hypothalamus to suppress food intake. Glucose-excited and glucose-inhibited neurons in the hypothalamus also participate in the regulation of appetite.

To a certain extent, the cerebral cortex can override peripheral signals to the hypothalamus for the regulation of appetite. For example, anorexia nervosa, a serious psychiatric disease, is characterized by excessive weight loss as a result of extreme dieting and physical activity. In comparison, excessive eating contributes significantly to the development of obesity in many countries. External factors—e.g., availability of favorite food in large portions at low cost—can often override satiety signals and lead to excessive eating and development of obesity.

KEY TERMS

- bile
- chief cells
- cholecystokinin (CCK)
- chylomicron
- D cells
- defecation
- diabetes mellitus
- digestion and absorption
- duodenum
- enterochromaffin-like cells
- external anal sphincter

- fat metabolism
- gastric accommodation
- gastric emptying
- gastrin
- gastrointestinal and hepatic circulation
- gastrointestinal motility
- gastrointestinal secretions
- G cells
- glucagon
- glucose homeostasis

- GLUT4 glucose carrier
- H^+-K^+-ATPase
- HCO_3^-/Cl^- exchanger
- hormone-sensitive lipase (HSL)
- ileocecal valve
- insulin
- internal anal sphincter
- lower esophageal sphincter
- pancreatic secretions
- parietal cells
- peristalsis

- ◆ peristaltic reflex
- ◆ peristaltic reflex
- ◆ pyloric sphincter

- ◆ satiety
- ◆ secretin
- ◆ somatostatin

- ◆ swallowing reflex
- ◆ upper esophageal sphincter

IMAGE CREDITS

REPRODUCTIVE PHYSIOLOGY

FETAL SEXUAL DIFFERENTIATION

Sex may be defined based on chromosomes, gonads, external genitalia, and **gender identity**, as shown in Fig. 10.1A. **Chromosomal sex** is probably the simplest definition of sex. The presence of Y chromosome is considered chromosomal male, whereas the absence of Y chromosome is considered chromosomal female. **Gonadal sex** is defined in terms of the presence of ovaries as female and the presence of testes as male. **Phenotypic sex**—appearance of external genitalia—is most commonly used for the definition of sex. For example, a newborn having a **penis** is considered a boy, whereas a newborn having a **vagina** is considered a girl. Phenotypic sex is typically consistent with gonadal sex in normal sexual differentiation. That is, the presence of a vagina is typically accompanied by the presence of ovaries, and the presence of a penis is typically accompanied by the presence of testes. Phenotypic sex can be inconsistent with gonadal sex in abnormal sexual differentiation. For example, patients having **complete androgen insensitivity syndrome (CAIS)** exhibit male gonadal sex in having testes but female phenotypic sex in having a vagina. The reason for the inconsistency is that CAIS patients do not have androgen receptors

LEARNING OBJECTIVES

1. **Fetal Sexual Differentiation.** Discuss the molecular mechanisms of fetal sexual differentiation; compare and contrast chromosomal sex, gonadal sex, phenotypic sex, and gender identify in terms of defining characteristics and regulatory mechanisms.

2. **Male Reproductive System.** Describe the anatomy of the male reproductive system, the regulation of spermatogenesis by sertoli cells, the regulation of reproductive hormone production by two negative feedback loops, and the regulation of erectile function by endothelial and neural cells.

3. **Female Reproductive System.** Describe the anatomy of the female reproductive system; identify the phases of the menstrual cycle; discuss the regulation of follicular development, ovulation, reproductive hormone production, and endometrial development by negative feedback and positive feedback during the menstrual cycle.

4. **Pregnancy and Parturition, and Lactation.** Discuss the physiology of pregnancy, including fertilization, embryo implantation, and placental development; discuss the mechanism of breast development during pregnancy; discuss the regulation of parturition by positive feedback; discuss the regulation of lactation by suckling reflex.

5. **Fetal Circulation.** Compare and contrast fetal and adult circulation in terms of anatomy and function.

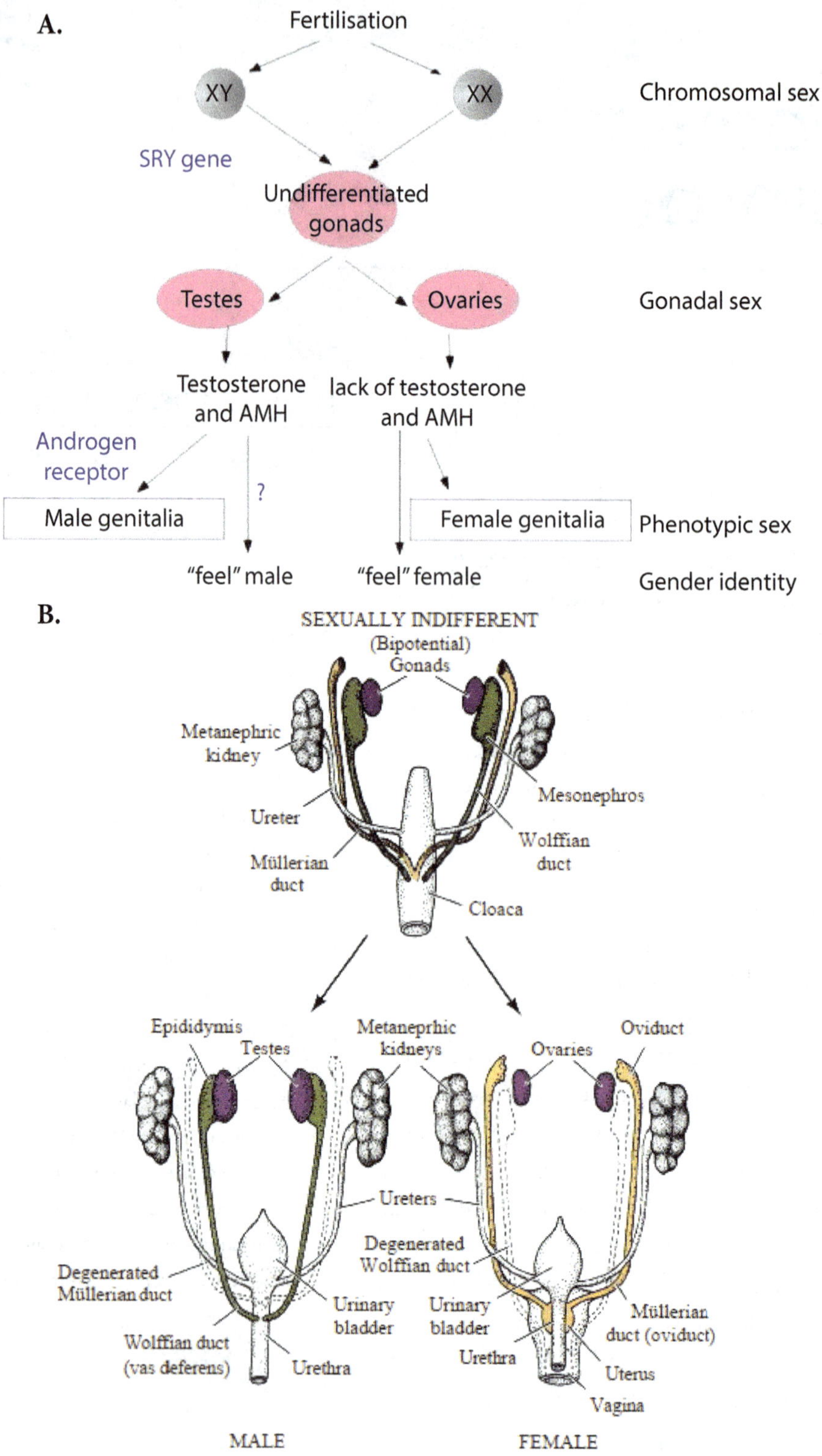

A.

B.

Fig. 10.1—A. Schematic diagram of fetal sexual differentiation. **B.** Sex-dependent fetal development of reproductive ducts.

for cells to respond to the male reproductive hormone—**testosterone**—released by the testes. Gender identity is the basic sense of being female or male, independent of any other definition of sex. Gender identity is most likely to be regulated by brain development.

General Principle of Fetal Sexual Differentiation.   As shown in Fig. 10.1A, the **sex-determining region of Y chromosome (SRY)** gene plays a primary role in directing the differentiation of fetal undifferentiated gonads into testes. Specifically, the SRY gene encodes the SRY transcription factor that is necessary for the development of testes. In the absence of the SRY gene, the default for fetal sexual differentiation is female. For example, patients with Klinefelter syndrome (47, XXY) are males, because the SRY gene enables the development of testes. In comparison, patients with Turner syndrome (45, XO) are females, because, in the absence of the SRY gene, the default for fetal sexual differentiation is the development of the female genitalia. In addition to the SRY gene, sex hormones and receptors are necessary for the progression of male sexual differentiation from testes to male genitalia. Genetic deficiencies in sex hormone-regulating enzymes—for example, 5α-reductase—and sex hormone receptors—for example, androgen receptors—can result in abnormal male phenotypic sex despite normal male chromosomal sex.

Male Fetal Sexual Differentiation.   As shown in Figs. 10.1A, in fetuses having the male chromosomal sex (46, XY), the presence of the SRY gene enables the undifferentiated gonads to develop into testes. Two hormones produced by the fetal testes—**anti-Mullerian hormone (AMH)**, also known as Mullerian inhibiting substance) and testosterone—are essential for the development of male genitalia. As shown in Fig. 10.1B, the hormone, anti-Mullerian hormone, produced by **Sertoli** cells in the testes, induces regression of the Mullerian ducts in the male fetus. The male sex hormone testosterone, produced by **Leydig cells** in the testes, enables the development of Wolffian ducts into vas deferens, epididymis, and seminal vesicles. Furthermore, the enzyme **5α-reductase** catalyzes the conversion of testosterone to **dihydrotestosterone (DHT)**, a hormone necessary for normal development of penis.

As shown in Fig. 10.1A, testosterone directs fetal development of the male genitalia. One implication of this mechanism is that exposure of female fetuses to testosterone, in utero, can lead to abnormal differentiation toward the male phenotype. Dihydrotestosterone is necessary for normal development of the penis. One implication of this mechanism is that exposure of male fetuses to 5α-reductase inhibitors, in utero, can lead to underdevelopment of the penis.

Female Fetal Sexual Differentiation.   Female chromosomal sex, that is, absence of Y chromosome, generally predicts female gonadal sex and female phenotypic sex, because the default of fetal sexual differentiation, in the absence of the SRY gene, is female. As shown in Fig. 10.1A, in fetuses having female chromosomal sex (46, XX), in the absence of the SRY gene, the undifferentiated gonads differentiate into ovaries. As shown in Fig. 10.1B, in the absence of testosterone, the Wolffian duct system regresses, and the Mullerian duct system develops into the oviduct (fallopian tubule) and uterus. In the absence of dihydrotestosterone, the primordial external genitalia develop into a vagina.

Meiosis—Basic Mechanism of Spermatogenesis and Oogenesis.   Meiosis is the basic cellular process for the production of sperms in males and ova in females. As shown in Fig. 10.2, meiosis begins with DNA replication, resulting in the formation of two sister chromatids per chromosome. During the prophase prior to meiosis I, crossover between homologous chromosomes can lead to the formation of recombinant chromatids from the two parent chromatids. Crossover is important for increasing the genetic diversity of sperms and eggs. During human meiosis I, homologous chromosomes segregate from each other to form two daughter cells having 23 chromosomes, with two chromatids per chromosome. During human meiosis II, the two chromatids of each chromosome separate from each other to form two daughter cells having 23 chromosomes, with one chromatid per chromosome.

Spermatogenesis by Meiosis. Fig. 10.3 shows the process of spermatogenesis by meiosis. Human cells contain 23 pairs of chromosomes. Fig. 10.3 illustrates chromosomal separation during spermatogenesis using only one pair of chromosomes. The first row in Fig. 10.3 shows a primary spermatocyte having 23 pairs of homologous chromosomes with two chromatids per chromosome. The different colored segments of chromatids in the primary spermatocyte represent the exchange of chromatid segments by crossover between homologous chromatids during the prophase before meiosis I. The second row in Fig. 10.3 shows the formation of two secondary spermatocytes after meiosis I. Each secondary spermatocyte contains 23 chromosomes, with two chromatids per chromosome. The third row in Fig. 10.3 shows the formation of four spermatids after meiosis II. Each spermatid contains 23 chromosomes, with one chromatid per chromosome. The fourth row in Fig. 10.3 shows the maturation of spermatids to fully differentiated spermatozoa.

MALE REPRODUCTIVE SYSTEM

Fig. 10.4A shows the anatomy of the male reproductive system, consisting of a penis, accessory glands (seminal vesicles, prostate gland), a pair of vas deferens, and two testes—organs of spermatogenesis. The descent of testes out of the abdominal cavity through the inguinal canal during fetal development is essential for male fertility, because spermatogenesis requires a temperature that is 2–4°C below core

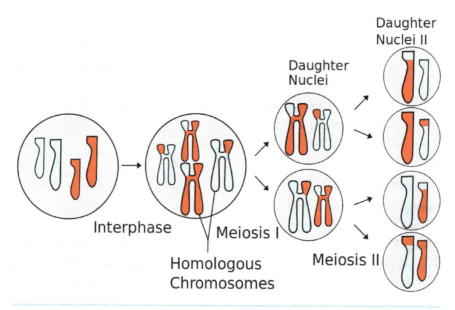

Fig. 10.2. Meiosis for the production of sperms and eggs.

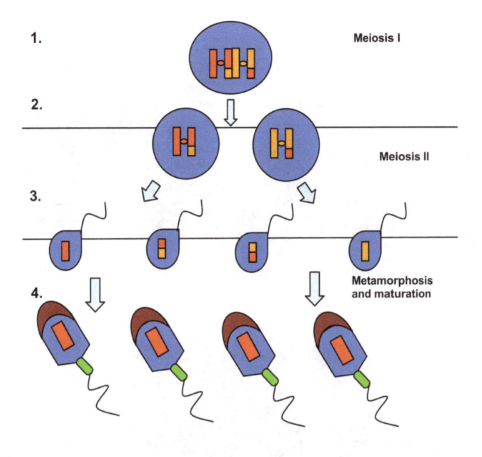

Fig. 10.3—Spermatogenesis by Meiosis in the Male Reproductive System. Labels: 1. primary spermatocyte; 2. secondary spermatocyte; 3. Spermatid; 4. Spermatozoa.

body temperature. Fig. 10.4B shows longitudinal and cross-sectional sections of a testis, consisting of seminiferous tubules for the production of sperms, epididymis for the storage of sperms, and vas deferens for the transportation of sperms to the penis.

Spermatogenesis. Fig. 10.5A shows the nurturing of spermatogenesis from undifferentiated spermatogonium to the differentiated spermatozoa by Sertoli cells in the seminiferous tubules of testes. As shown in Fig. 10.5A, the **blood-testis barrier** separates the Sertoli cell-formed seminiferous epithelium into the outer basal and inner adluminal compartments. The outer basal compartment is the location of spermatogonial renewal and differentiation. The inner adluminal compartment is the location of the production of secondary spermatocytes from primary spermatocytes by meiosis I and II, and differentiation of spermatocyte into spermatid. The blood-testis barrier is dynamic and constantly undergoing periodic restructuring to allow the crossing of spermatocyte from the basal compartment to the adluminal compartment.

Fig. 10.5B shows a fully differentiated sperm, consisting of a head, mid-piece, a tail, and end piece. The head of a sperm holds two important structures—nucleus and acrosome. The nucleus stores the

A.

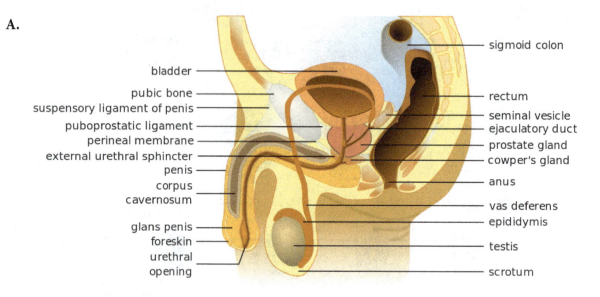

sigmoid colon

bladder

pubic bone

suspensory ligament of penis

puboprostatic ligament

perineal membrane

external urethral sphincter

penis

corpus
cavernosum

glans penis

foreskin

urethral
opening

rectum

seminal vesicle
ejaculatory duct

prostate gland

cowper's gland

anus

vas deferens

epididymis

testis

scrotum

B.

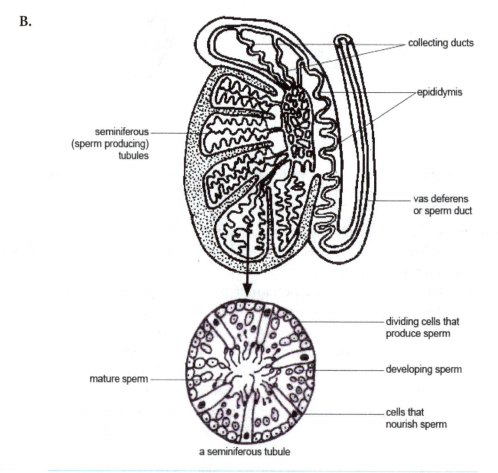

collecting ducts

epididymis

seminiferous
(sperm producing)
tubules

vas deferens
or sperm duct

dividing cells that
produce sperm

developing sperm

mature sperm

cells that
nourish sperm

a seminiferous tubule

Fig. 10.4—A. Male reproductive system   **B.** Structure of a testis.

genetic material, whereas the acrosome contains enzymes that facilitate the entry of a sperm into an ovum during fertilization. The mid-piece contains mitochondria for energy metabolism. The tail contains motor proteins for sperm motility. Semen ejaculated by the penis during intercourse typically consists of approximately 10 percent sperms and 90 percent seminal plasma—secretions from seminal vesicles and the prostate gland.

Male Reproductive Endocrine System. Fig. 10.6 shows the two distinct branches of the male reproductive hormonal system for regulation of spermatogenesis and secretion of the male reproductive hormone, testosterone. As shown in Fig. 10.6, the hypothalamus secretes a gonadotropin-releasing hormone (GnRH), which stimulates the anterior pituitary gland to release two hormones—follicle-stimulating

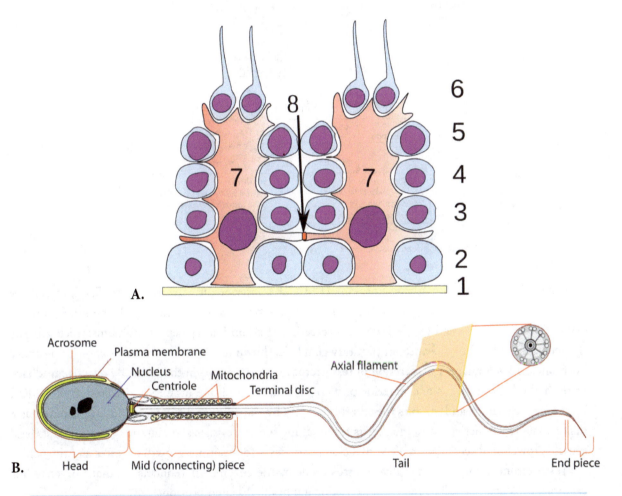

Fig. 10.5—A. Schematic diagram of sertoli cells and their function spermatogenesis. 1. Basal lamina; 2. Spermatogonia; 3. Primary spermatocyte; 4. Secondary spermatocyte; 5. Spermatid; 6. Spermatozoa; 7. Sertoli Cell; 8. Tight junction of blood-testis barrier. **B.** Structure of a human spermatozoa.

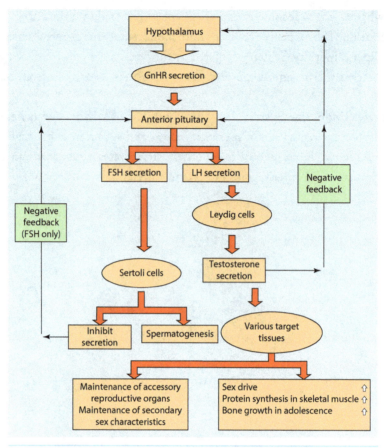

Fig. 10.6. Male reproductive hormonal system.

hormone (FSH) and a luteinizing hormone (LH). The anterior pituitary hormone, FSH, stimulates spermatogenesis as nurtured by Sertoli cells in the testis. By negative feedback to the anterior pituitary gland, Sertoli cells respond to FSH with the secretion of an inhibitory peptide, inhibin, which inhibits the secretion of FSH by the anterior pituitary gland. As shown in Fig. 10.6, the other anterior pituitary hormone, LH, stimulates the secretion of testosterone by Leydig cells in the testis. By negative feedback to both the hypothalamus and anterior pituitary gland, testosterone inhibits the secretion of GnRH by the hypothalamus and inhibits the secretion of LH by the anterior pituitary gland. Testosterone is a major male reproductive hormone that is essential for spermatogenesis, maintaining sex drive, skeletal muscle growth, and secondary male sexual characteristics—for example, growth of testes and the penis.

Testosterone is converted to dihydrotestosterone by the enzyme 5α-reductase in many organs—for example, prostate, testis, hair follicles, liver, skin, and brain. As cited previously (Fig. 10.1), dihydrotestosterone, during male fetal development, is essential for the development of external genitalia. In adults, dihydrotestosterone is the major androgen in the prostate gland. Excessive production of dihydrotestosterone catalyzed by 5α-reductase in the prostate can cause abnormal enlargement of the prostate (benign prostate hyperplasia) and obstruction of the urinary tract. Pharmacologic inhibitors of 5α-reductase are available for the treatment of benign prostate hyperplasia.

Erectile Function. Penile erection is a complex spinal reflex that can be initiated by mechanical, visual, and mental stimuli. The erectile tissue of a penis is the vascular tissue, known as **corpora cavernosa**, consisting of vascular sinusoids lined with endothelial and vascular smooth muscle cells. When the corpora cavernosa is filled with blood, the penis becomes erect. Development of penile erection is mediated by the following vascular events: a) dilation of cavernosal arterioles and arteries increases the inflow of blood to the corpora cavernosa; b) relaxation of cavernosal smooth muscle cells increases the compliance of corpora cavernosa for filling of blood; and c) compression of cavernosal veins by the erectile tissue reduces the outflow of blood from corpora cavernosa, thereby maintaining rigidity of the penis.

As shown in Fig. 10.7, **nitric oxide (NO)** released by neurons (nonadrenergic and noncholinergic) and endothelial cells and prostaglandin E_1 released by cells within the penis are the major mediators of cavernosal smooth muscle relaxation during penile erection. NO is synthesized from L-arginine by **neuronal nitric oxide synthase (nNOS)** in neurons and **endothelial NO synthase (eNOS)** in endothelial cells. Both nNOS and eNOS are stimulated by intracellular $[Ca^{2+}]$, as summarized in the following scheme:

$$\uparrow \text{Intracellular } [Ca^{2+}] \rightarrow \text{Activation of nNOS or eNOS} \rightarrow \text{Synthesis of NO from L-Arginine}$$

Electrical stimulation leads to an increase in intracellular $[Ca^{2+}]$ in nonadrenergic, noncholinergic neurons, whereas parasympathetic stimulation via muscarinic acetylcholine receptors leads to an increase in intracellular $[Ca^{2+}]$ in endothelial cells. As shown in Fig. 10.7, the lipid-soluble molecule NO diffuses into cavernosal smooth muscle cells to induce muscle relaxation by activating the following cascade of enzymatic reactions:

1. NO activates guanylyl cyclase

2. Guanylyl cyclase catalyzes the formation of cyclic GMP (cGMP) from GTP

3. cGMP activates cGMP-dependent protein kinase (cGK)

4. cGK stimulates Ca^{2+} uptake by the sarcoplasmic reticulum, inhibits calcium channels, and activates potassium channels on the cell membrane

5. Decrease in intracellular $[Ca^{2+}]$ in cavernosal smooth muscle cells

6. Relaxation of cavernosal smooth muscle cells

7. Penile erection

cGMP is the intracellular messenger that mediates cavernosal smooth muscle relaxation for penile erection. As shown in Fig. 10.7, guanylyl cyclase catalyzes the formation of cGMP from GTP, whereas phosphodiesterase 5 (PDE5) catalyzes the degradation of cGMP to GMP. Accordingly, the guanylyl cyclase/PDE5 activity ratio is a major determinant of intracellular [cGMP] in cavernosal smooth muscle cells and the strength of penile erection. Based on this knowledge, pharmacologic inhibitors of PDE5

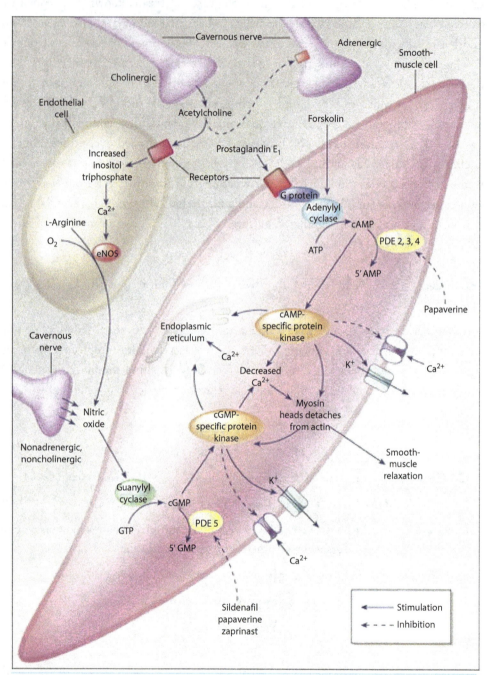

Fig. 10.7—Molecular Mechanism of Penile Smooth Muscle Relaxation for Penile Ejection. From: Dean
RC and TF Lue. Urol Clin N Am 32: 379–395, 2005.

have been developed for the treatment of erectile dysfunction by enhancing intracellular [cGMP] in cavernosal smooth muscle cells.

Another important mediator of penile erection is prostaglandin E_1 (PGE_1), released by cells within the penis. PGE_1 induces cavernosal smooth muscle relaxation by activating a G protein-coupled receptor on the cell membrane, thereby triggering the following cascade of enzymatic reactions: a) activation of adenylyl cyclase; b) conversion of ATP to cyclic AMP (cAMP); c) activation of cAMP-dependent protein kinase (PKA); d) PKA-mediated stimulation of Ca^{2+} uptake by the sarcoplasmic reticulum; inhibition of calcium channels and activation of potassium channels on the cell membrane; e) decrease in intracellular $[Ca^{2+}]$; f) cavernosal smooth muscle relaxation; and g) penile erection. Cyclic AMP is degraded to AMP by phosphodiesterase 2, 3, 4 (PDE 2, 3, 4) in cavernosal smooth muscle cells. Based on this knowledge, PGE_1 injection and pharmacologic inhibitors of PDE 2, 3, 4 have been developed for the treatment of erectile dysfunction.

FEMALE REPRODUCTIVE SYSTEM

Fig. 10.8 shows the female reproductive system, consisting of the clitoris, labium minus and labium majus, vagina, cervix, uterus, uterine (fallopian) tubes, and ovaries. The clitoris, the main organ of orgasm in females, is similar to the penis in being developed from the same embryological structure and having cavernous erectile tissue. The vagina is the entrance to and exit from the female reproductive tract. The

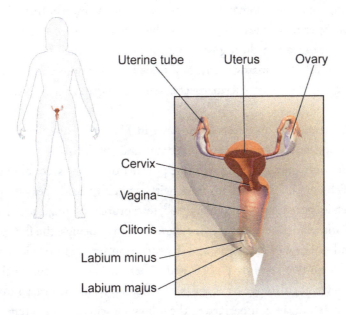

The Female Reproductive System

Fig. 10.8. Female reproductive system.

vaginal mucosa is responsive to female reproductive hormones and undergoes changes in thickness during a menstrual cycle. In nonpregnant females, the vaginal mucosal layer reaches its peak thickness at approximately the middle of the menstrual cycle. The uterus is the site of embryo implantation and placental formation, and fetal development. Uterine (fallopian) tubes are conduits for the transport of ova from ovaries toward the uterus for fertilization with a sperm, which typically occurs inside the fallopian tube.

Oogenesis is similar to spermatogenesis in utilizing meiosis for the production of ovum, but is significantly different from spermatogenesis in several aspects. Fig. 10.9 shows the beginning of oogenesis at the primary oocyte stage, because all oogonia have developed into primary oocytes by the time of birth. One implication of this characteristic of oogenesis is that an oocyte released at the time of ovulation is as old as the ovulating person. Errors in meiosis tend to increase with age, and maternal age

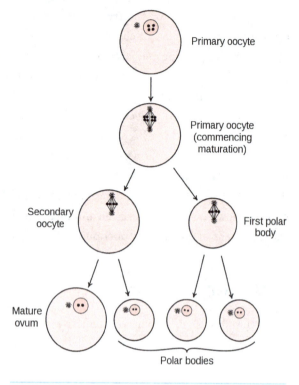

Fig. 10.9. Oogenesis.

is a risk factor for developing chromosomal abnormalities in the ovum. Another distinguishing characteristic of oogenesis from spermatogenesis is that typically only one primary oocyte becomes dominant for development into a mature ovum during each menstrual cycle, which lasts approximately twenty-eight days. Furthermore, unlike spermatogenesis, oogenesis is an asymmetric process. Spermatogenesis produces four sperms from one primary spermatocyte, whereas oogenesis produces only one ovum from a primary oocyte.

Fig. 10.9 shows the process of oogenesis. As shown in Fig. 10.9, one primary oocyte undergoes maturation with DNA duplication, resulting in 46 chromosomes, with two chromatids per chromosome. The primary oocyte then undergoes meiosis I with the production of one secondary oocyte and the first polar body, both of which contain 23 chromosomes, with two chromatids per chromosome. Meiosis I in oogenesis is asymmetric, because the secondary oocyte captures the majority of the cytoplasm and will undergo meiosis II with the production of an ovum. In human beings, the first polar body contains very little cytoplasm and undergoes extrusion and degeneration. In some species, the first polar body undergoes meiosis II to form two secondary polar bodies before degeneration. Furthermore, in human beings, the secondary oocyte at this stage of oogenesis is released from the ovary into the fallopian tube by a process known as ovulation, as summarized in the following schematic diagram:

Ovulation → Secondary Oocyte + First Polar Body

After ovulation, the secondary oocyte migrates down the fallopian tube toward the uterus for fertilization by a sperm. If fertilization does not occur, the secondary oocyte will not undergo meiosis II and be expelled by the uterus during a menstrual cycle. If fertilization occurs, the secondary oocyte will then undergo meiosis II with the production of one mature ovum and the second polar body, both of which contain 23 chromosomes, with one chromatid per chromosome. The mature ovum fuses with the sperm to form a zygote, whereas the second polar body is extruded, as summarized in the following schematic diagram:

Fertilization → Meiosis II → Mature Ovum + Second Polar Body

Mature Ovum + Sperm → Zygote

Female Reproductive Endocrine System. Reproductive hormones in human females undergo significant oscillations during a menstrual cycle, because female reproductive endocrine cells and the oocyte are parts of the ovarian follicle that undergo development and degeneration during a menstrual cycle. As shown in Fig. 10.10, an ovarian follicle consists of an oocyte and follicular fluid in the center, with outer layers of granulosa and theca cells. Granulosa cells serve two functions: nurturing of ovarian follicular development, and conversion of androgens to estrogens. Theca cells are endocrine cells that secrete androgens (androstenedione and testosterone), which are then converted by the enzyme aromatase in granulosa cells to estrogens (estrone and estradiol), as summarized in the following schematic diagram:

Theca Cells → Androgens → Granulosa Cells → Estrogens

The ovarian follicle undergoes development and degeneration during a menstrual cycle. As part of the ovarian follicle, theca cells and granulosa cells also undergo development and degeneration during menstrual cycles, resulting in significant fluctuations in plasma concentrations of reproductive hormones in females. The endometrial layer of the uterus also undergoes growth and degeneration during a menstrual cycle, because endometrial growth is dependent on reproductive hormones.

Fig. 10.11 shows the female reproductive hormonal system, which is similar to the male reproductive hormonal system in being regulated by the same hypothalamic hormone (GnRH, also known as luteinizing-releasing hormone, LHRH) and the same anterior pituitary hormones (FSH and LH). During ovarian follicular development, FSH stimulates granulosa cells to secrete factors that are

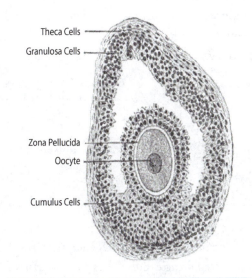

Fig. 10.10—Structure of an ovarian follicle.

essential for ovarian follicular development. In addition, FSH stimulates granulosa cells to convert androgens to estrogens for secretion. By negative feedback to the anterior pituitary gland, granulosa cells respond to FSH with the secretion of the inhibitory peptide inhibin, which inhibits the secretion of FSH by the anterior pituitary gland. LH stimulates theca cells to secrete androgens, which are converted by granulosa cells to estrogens. By negative feedback to both the hypothalamus and anterior pituitary gland, estrogens inhibit the secretion of GnRH by the hypothalamus and the secretion of LH by the anterior pituitary gland.

After ovulation, the postovulatory ovarian follicle is transformed into **corpus luteum**—an endocrine gland that secretes **estrogen** and **progesterone** in response to stimulation by LH, as shown in Fig. 10.11. Progesterone secreted by corpus luteum is essential for maintaining the uterine **endometrium** for **implantation** of the embryo during the early stage of pregnancy. By negative feedback to both the hypothalamus and anterior pituitary gland, estrogen and progesterone inhibit the secretion of GnRH by hypothalamic cells and secretion of LH by anterior pituitary cells.

Menstrual Cycle. The menstrual cycle refers to the cyclical changes in the female reproductive system—development and degeneration of ovarian follicle, fluctuations in the secretion

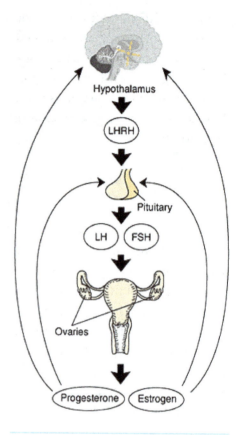

Fig. 10.11—Female Reproductive Hormonal System. From: National Institutes of Health

of reproductive hormones, and growth and degeneration of the uterine endometrium. The duration of a menstrual cycle is approximately twenty-eight days. The menstrual cycle may be understood by focusing on the following sequence of changes in the female reproductive system:

Follicular Development and Degeneration

↓

Fluctuations in the Secretion of Reproductive Hormones

↓

Uterine Endometrial Growth and Degeneration

Uterine endometrial degeneration, resulting in menstrual bleeding, is the most noticeable event in a menstrual cycle and marks the first day of the menstrual cycle. As shown in Fig. 10.12 (**fourth row, "Endometrial Histology"**), the first day of a menstrual cycle is marked by the beginning of endometrial degeneration, when the levels of estrogen and progesterone have fallen to their lowest levels

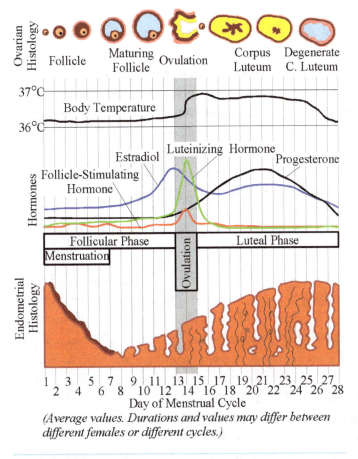

Fig. 10.12. Menstrual cycle.

after degeneration of the corpus luteum. The degenerated endometrium is expelled by the uterus as menstrual blood.

The Follicular Phase refers to development of the ovarian follicle from the first day of the menstrual cycle to the day before ovulation, as shown in Fig. 10.12 (**first row, "Ovarian Histology"**). Body temperature during the follicular phase is approximately 1°C lower than the body temperature after ovulation, as shown in Fig. 10.12 (**second row, "Body Temperature"**). Change in body temperature, in combination with the number of days after the onset of menstrual bleeding, is sometimes used to determine the time of ovulation and fertile window for conception. Plasma concentrations of FSH and LH increase during the follicular phase, because GnRH-secreting cells in the hypothalamus and FSH and LH-secreting cells in the anterior pituitary become active in the absence of feedback inhibition by estrogen and inhibin, as shown in Fig. 10.12 (**third row, "Hormones"**). FSH and LH together stimulate development of the oocyte, theca cells, and granulosa cells in the ovarian follicle. Theca cells secrete androgens, which are converted to estrogens by granulosa cells. Plasma concentration of estrogen (estradiol) increases during the follicular phase and reaches a peak level just before ovulation, as shown in Fig. 10.12 (**third row, "Hormones"**). Plasma concentration of androgens also increases during the follicular phase. In response to the rising level of estrogen, the uterine endometrium begins to develop on approximately the eighth day of the menstrual cycle, as shown in Fig. 10.12 (**fourth row, "Endometrial Histology"**).

In summary, the follicular phase is characterized by follicular development, increases in plasma concentrations of FSH, LH, estrogens and androgens, and uterine endometrial development.

Ovulation is the process of releasing an ovum from an ovarian follicle, which typically occurs in the middle of a menstrual cycle—that is, day fourteen of a twenty-eight-day cycle. The molecular mechanism of ovulation is not completely understood, but a positive feedback between estrogen and LH leading to a surge in LH is generally recognized as an important trigger of ovulation, as shown in Fig. 10.12 (**third row, "Hormones"**) and summarized in the following scheme:

$$\text{High Plasma [Estrogen]} \rightarrow \text{GnRH Surge} \rightarrow \text{LH Surge} \rightarrow \text{Ovulation}$$

Estrogen-induced LH surge is considered positive feedback, because LH is a stimulant of estrogen secretion. The positive feedback between estrogen-stimulation LH secretion and LH-stimulated estrogen secretion will amplify the secretion of LH and estrogen until both hormones surge to very high levels. A surge in FSH also occurs immediately before ovulation. The LH surge eventually stimulates ovulation—the release of a secondary oocyte from the ovarian follicle, as shown in Fig. 10.12 (**first row, "Ovarian Histology"**). The released secondary oocyte is normally captured by the fallopian tube and transported by ciliated cells along the tube toward the uterus. The unfertilized secondary oocyte typically survives in the fallopian tube for up to two days, which is considered the "fertile window" for conception.

After ovulation, the postovulatory ovarian follicle begins to evolve into a different reproductive endocrine gland, the corpus luteum, and the menstrual cycle enters the luteal phase.

Luteal Phase refers to the development of the corpus luteum—an endocrine gland evolved from the postovulatory ovarian follicle, as shown in Fig. 10.12 (**first row, "Ovarian Histology"**). The term "corpus luteum" means yellow body, because the structure is yellow in color. An important function of corpus luteum is the secretion of progesterone and estrogen, essential hormones for supporting the growth of the uterine endometrium for embryo implantation, as shown in Fig. 10.12 (**third panel, "Hormones" and fourth row, "Endometrial Histology"**). If the secondary oocyte is not fertilized by a sperm, the corpus luteum will spontaneously undergo degeneration, resulting in the fall in plasma concentration of estrogen and progesterone, degeneration of the uterine endometrium, and menstrual bleeding. If the secondary oocyte is fertilized by a sperm, and the embryo is implanted into the uterine endometrium, the placenta-forming cells will secrete human chorionic gonadotropin (hCG) that rescues the corpus luteum from degeneration. The corpus luteum will then continue to secrete progesterone for four to five weeks after implantation of the embryo, by which time the placenta becomes the major source of estrogen and progesterone for sustaining pregnancy.

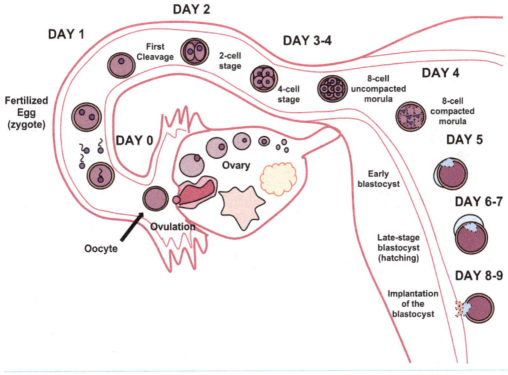

Fig. 10.13. Fertilization to implantation.

PREGNANCY AND PARTURITION

Fertilization and Embryo Implantation. Sperm motility is essential for fertilization, because fertilization between an oocyte and a sperm typically occurs in the fallopian tube at one to two days after ovulation, as shown in Fig. 10.13. The fertilized egg (zygote) then undergoes cell divisions to form an embryo while migrating down the fallopian tube toward the uterus for implantation onto the uterine endometrium. Implantation of the embryo at the blastocyst stage typically occurs at eight to nine days after fertilization.

Fig. 10.14A shows the process of fertilization of an oocyte by a sperm. A sperm undergoes capacitation during its migration through the female reproductive tract to become competent for fertilization. Sperm capacitation consists of releasing cholesterol, glycoprotein, and other molecules from the surface of the sperm, thereby enabling the acrosome at the head of a sperm to undergo the necessary acrosomal reaction for entering the oocyte. As shown in Fig. 10.14A, a sperm penetrates the matrix of cumulus cells and undergoes the acrosomal reaction, in which acrosomal enzymes are released by exocytosis. An acrosome-reacted sperm then binds to the **zona pellucida** of the ovum. Acrosomal enzyme-mediated proteolysis of the zona pellucida (ZP) and motility of the sperm flagella together enable the penetration of the sperm through the zona pellucida into the perivitelline space (PVS), and fusion of the sperm with the oocyte membrane via specific oocyte surface proteins for entry into the oocyte. Fertilization of the oocyte triggers meiosis II with the formation of a mature ovum and the secondary polar body, as

A.

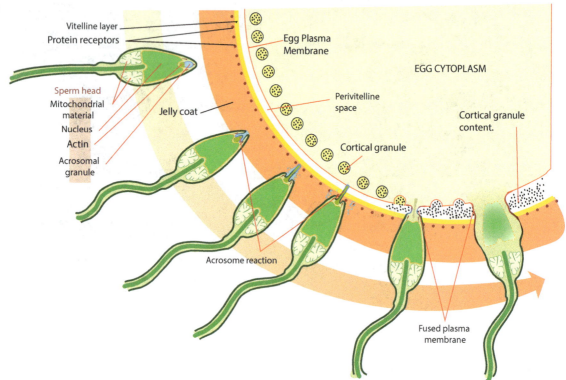

Vitelline layer
Protein receptors
Egg Plasma Membrane
EGG CYTOPLASM
Sperm head
Mitochondrial material
Perivitelline space
Cortical granule content.
Nucleus
Jelly coat
Actin
Acrosomal granule
Cortical granule
Acrosome reaction
Fused plasma membrane

B.

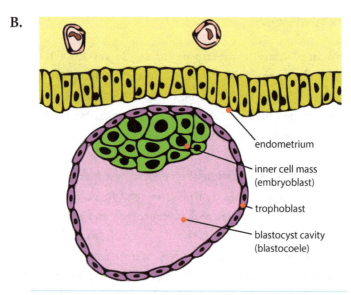

endometrium

inner cell mass
(embryoblast)

trophoblast

blastocyst cavity
(blastocoele)

Fig. 10.14—A. Fertilization. **B.** Structure of blastocyst before implantation.

examined previously. Fusion of the mature ovum with sperm leads to the formation of zygote, which then undergoes cell division to form an embryo. Polyspermy is blocked by the following mechanisms. Upon fertilization, the fertilized ovum releases cortical granules from the cortex by exocytosis, which causes the zona pellucida to become non-receptive to binding of other sperms.

Fig. 10.14B shows the structure of a blastocyst at the time of implantation, which consists of an inner cell mass (embryoblast), cavity (blastocoele), and the surrounding trophoblast. Contact between the blastocyst and uterine endometrium stimulate endometrial and trophoblast cells to undergo proliferation and differentiation for the formation of placenta. As shown in Fig. 10.15A, the placenta functions as an interface for diffusional transport of nutrients and waste between maternal and fetal circulation. Fig. 10.15B shows the structure of a placenta. The chorionic villi, fingerlike structures within the placenta, are formed by the invasion of trophoblast cells from the embryo. Chorionic villi are perfused by umbilical vessels of the fetus, whereas the intervillous space of the placenta is perfused by maternal blood vessels.

The placenta also functions as an important endocrine gland. During embryo implantation, the placenta secretes human chorionic gonadotropin (hCG), an essential hormone for rescuing the corpus luteum from degeneration. Human chorionic gonadotropin can be measured in maternal blood as early as forty-eight hours after implantation, and appearance of hCG in maternal blood is an indicator of pregnancy. Secretion of progesterone by the corpus luteum is essential for maintaining the uterine endometrium for placental development during early pregnancy. After the sixth to tenth week of pregnancy, the placenta becomes the major source of progesterone and estrogen. Progesterone is important for maintaining quiescence of the myometrium for fetal growth during pregnancy. Estrogen is important for supporting the development of uterine myometrial contractility near term, breast development during pregnancy, and cervical ripening during labor.

Parturition (Labor) is a highly complex process that is regulated by hormones, inflammatory cytokines, and local mediators. Time-dependent changes in uterine contractility and cervical opening are essential for successful pregnancy and parturition. Uterine smooth muscle cells are quiescent during pregnancy for accommodating fetal growth, but become contractile near term for expelling the baby during parturition. The cervix is rigid and closed during pregnancy for bearing the weight of the fetus, but becomes soften and ripened (loss of tissue compliance) near term for allowing the baby to exit the uterus during parturition. Timing of parturition remains poorly understood. The progesterone/estrogen activity ratio appears to be an important determinant of uterine smooth muscle contractility and cervical opening during pregnancy. A high progesterone/estrogen activity ratio during pregnancy inhibits uterine smooth muscle contractility and cervical ripening, thereby accommodating and holding the growing fetus. Progesterone inhibits uterine contractility in part by suppressing the expression of oxytocin receptors in uterine smooth muscle cells. A reversal of progesterone/estrogen activity ratio (progesterone withdrawal) near term results in the development of uterine smooth muscle contractility and cervical ripening. Treatment with progesterone has been used to prevent preterm birth.

The posterior pituitary hormone, oxytocin, stimulates uterine smooth muscle contractility during parturition. In a positive feedback, the contracting uterus sends neural signals to the brain to stimulate secretion of oxytocin by the posterior pituitary gland, thereby maximizing oxytocin release and uterine contractions for expelling the baby from the uterus. The placenta also produces a large amount of oxytocin

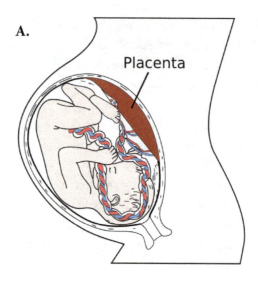

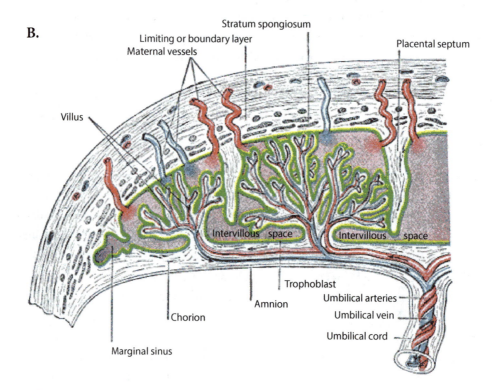

Fig. 10.15—A. Placenta for supporting fetal growth.   **B.** Diffusional transport of nutrient and waste between maternal and fetal blood at the placenta.

Fetal Circulation

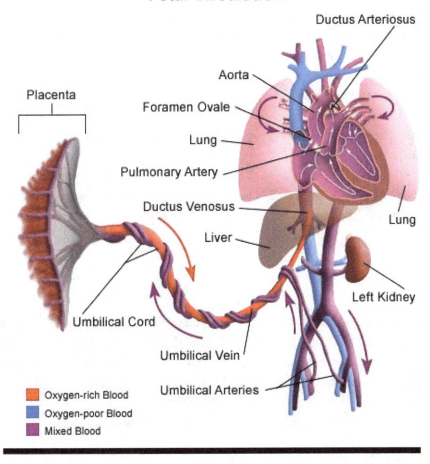

Fig. 10.16. Fetal circulation.

during pregnancy, but the contribution of placental oxytocin to parturition is unclear. Prostaglandins within the uterus stimulate uterine smooth muscle contractions near term. In addition, prostaglandins stimulate the degradation and reorganization of the extracellular matrix—for example, collagen, in the cervix, thereby causing cervical softening and ripening for the baby to exit the uterus during parturition. Oxytocin and prostaglandins E_1 and E_2 are sometimes used for the induction of labor.

FETAL CIRCULATION

Lungs in the fetus are not ventilated with air. The placenta functions as a gas exchanger between maternal circulation and fetal circulation. As shown in Fig. 10.16, umbilical arteries carry deoxygenated blood from the fetal systemic arterial system to the placenta for oxygenation, and the umbilical vein carries oxygenated blood from the placenta to the fetal systemic venous system via the ductus venosus. Due to the presence of high vascular resistance in the fetal pulmonary circulation and two low-resistance shunts between the right and left fetal heart, a significant amount of systemic venous return to the right heart flows directly into the fetal systemic arterial system, bypassing the pulmonary circulation.

The **foramen ovale** is an opening between the right and left atrium in the fetal heart that allows a significant amount of systemic venous return to the right atrium to flow into the left atrium, instead of the right ventricle, thereby reducing the amount of blood flow into the pulmonary circulation, as shown in Fig. 10.16 and the following scheme:

$$\text{Right Atrium} \rightarrow \text{Foramen Ovale} \rightarrow \text{Left Atrium}$$

The **ductus arteriosus** is a low-resistance vascular conduit between the main pulmonary artery and aorta in the fetal heart, which allows a substantial amount of pulmonary arterial blood to flow into the aorta, thereby reducing the amount of blood flow into the pulmonary circulation, as shown in Fig. 10.16 and the following scheme:

$$\text{Pulmonary Artery} \rightarrow \text{Ductus Arteriosus} \rightarrow \text{Aorta}$$

Overall, total pulmonary blood flow is less than total systemic blood flow in the fetal circulation.

LACTATION

During pregnancy, estrogen, progesterone, and **prolactin** stimulate breast development and maturation by stimulating the development of alveoli in the breast for milk secretion and storage, and the ductal system for carrying milk to the nipple. Prolactin is the major hormone for stimulating milk synthesis by the alveolar epithelial cells. The increase in plasma concentration of prolactin during pregnancy is largely due to stimulation by estrogen. Progesterone stimulates breast development but inhibits milk synthesis. Plasma progesterone level falls after expulsion of the placenta, thereby removing the inhibitory effect of progesterone on milk synthesis.

Immediately after parturition, the breast produces colostrum, a fluid containing low concentration of milk protein and high concentrations of immunoglobulin and other biological factors that stimulate the development of the gastrointestinal tract and immune system in the newborn. Milk synthesis begins at two to three days after parturition, but mature milk production occurs at two to three weeks after parturition. Suckling of the breast by an infant or pump stimulates secretion of the anterior pituitary hormone, prolactin, possibly by removing the inhibitory effect of dopamine on prolactin-secreting cells. Prolactin stimulates milk synthesis by the alveoli after parturition. In a milk-ejection reflex, suckling of the breast stimulates the secretion of the posterior pituitary hormone, **oxytocin**, which stimulates the contraction of myoepithelial cells surrounding the alveoli, resulting in milk ejection. Oxytocin also stimulates the uterine smooth muscle contractions, thereby facilitating return of the uterus to the pre-pregnancy size.

Weaning—cessation of breast-feeding—induces the loss of alveoli and return of the breast to the non-lactating state, but the underlying mechanism is not fully understood. Relactation is the process of restarting lactation after a period of weaning by mechanical stimulation of the breast with a breast pump and/or nursing child. Mechanical stimulation of the breast appears to be sufficient to stimulate the release of prolactin and induce lactation, because it is possible to induce lactation in nonpregnant women by mechanical stimulation of the breast using a breast pump. Alternatively, dopamine receptor antagonists can also be used to induce lactation by blocking the inhibitory effect of dopamine on prolactin secretion. The process of lactation suppresses the release of gonadotropin-releasing hormone from the hypothalamus, thereby delaying the onset of menstrual cycles in breast-feeding women. In some countries, lactation has been proposed as a natural method of contraception.

KEY TERMS

- 5α-reductase
- anti-Mullerian hormone (AMH)
- blood-testis barrier
- chromosomal sex
- complete androgen insensitivity syndrome (CAIS)
- corpora cavernosa
- corpus luteum
- cortical reaction
- dihydrotestosterone (DHT)
- ductus arteriosus
- endometrium
- endothelial NO synthase (eNOS)
- erectile function
- estrogen
- fetal circulation
- fetal sexual differentiation
- foramen ovale
- gender identity
- gonadal sex
- granulosa cell
- guanylyl cyclase
- implantation
- lactation
- Leydig cell
- menstrual cycle
- neuronal nitric oxide synthase (nNOS)
- nitric oxide (NO)
- oogenesis
- ovulation
- oxytocin
- parturition
- parturition (labor)
- penis
- phenotypic sex
- polar body
- progesterone
- prolactin
- prostaglandin E$_1$
- sertoli cell
- sex-determining region of Y chromosome (SRY)
- spermatogenesis
- testosterone
- theca cell
- vagina
- zona pellucida

IMAGE CREDITS

INDEX

resting membrane potential, 18, 21, 21–22, 26, 73
restrictive lung disease, 118

S

saltatory conduction, 27
sarcomere, 34–36, 35, 38, 43, 43–44
sarcoplasmic reticulum, 24, 39–41, 53–56, 80, 193–194
satiety, 183
saturation kinetics, 10, 11–12
secretin, 168, 171–172
semen, 191
seminal vesicle, 187–188, 191
sensory neuron, 34, 52–53
sertoli cell, 187
set point, 99–100
sex-determining region of Y chromosome, 187
SGLT1, 174
SGLT2, 141, 181
shallow breathing, 121
sinoatrial node, 71–72, 87, 105–106, 106
skeletal muscle, 10, 17, 22, 24, 33–35, 37, 37–39, 43,
    44, 45, 53, 75, 80–81, 95, 102, 105, 157–158,
    166, 167, 178, 181, 192
slow oxidative fiber, 50
small intestine, 161–162, 166–173
smooth muscle, 33–34, 54–56, 65, 89, 95–98, 100,
    157–159, 166–167, 193–195, 203–204, 207
SNAP-25, 38
sodium channel, 25, 73
somatic nervous system, 157, 166
somatostatin, 64, 170, 171
spatial summation, 29, 29–30
sperm, 63, 187–190, 196–197, 200–202
spermatocyte, 188–190, 196
spermatogenesis, 63, 187–191, 196
spermatogonium, 189
spermatozoa, 188–190
sphincter of Oddi, 172
spinal cord, 24, 30–31, 34, 52–54, 156

spirometry, 110, 112
SRY, 187–188
standard ECG leads, 77
standard leads, 78–79
steroid hormone, 57, 97–98, 148
stomach, 161–162, 166, 168–169, 174
stretch-activated ion channel, 10
stretch reflex, 52–53
striated muscle, 33, 33–34, 34, 53, 54, 80
stroke volume, 82, 83, 87–88, 90–91, 105–107, 146
S-T segment, 75
substrate metabolism, 161–181
suckling, 64, 207
sucrase, 173
sucrose, 173
supra-threshold excitatory synapse, 29
surface tension, 113
surfactant, 118–119
swallowing reflex, 163, 164
sympathetic nervous system, 28, 30, 148, 167–168
sympathetic post-ganglionic fiber, 30
sympathetic pre-ganglionic fiber, 30–31
synaptic cleft, 13, 23–24, 29, 37
synaptic vesicle, 13, 23–24, 26, 38–39
synaptobrevin, 38
syntaxin, 38
systemic arterial pressure, 69
systemic circulation, 68–69, 69, 82, 84, 98–99, 105,
    124, 148
systole, 82, 83, 86
systolic pressure, 84–85, 85, 90–91

T

T3, 61–62
T4, 61–62
TAG, 181
temporal summation, 29, 29–30
testis, 189–191
testosterone, 57, 60, 63, 187, 187–188, 191–192, 197

CPSIA information can be obtained
at www.ICGtesting.com
Printed in the USA
LVHW060126080421
683743LV00002B/4